The Carpenter's Toolbox Manual

Gary D. Meers

ARCO

New York

ARCO

Simon & Schuster, Inc.
15 Columbus Circle
New York, NY 10023

DISTRIBUTED BY PRENTICE HALL TRADE SALES

Manufactured in the United States of America

   2   3   4   5   6   7   8   9   10

Library of Congress Cataloging-in-Publication Data

Meers, Gary D.
    The carpenter's toolbox manual / by Gary D. Meers ;
In Plain English, Inc.
        p.   cm.
    ISBN 0-13-115296-3
    1. Carpentry—Handbooks, manuals, etc.   I. In Plain
English, Inc.   II. Title.
TH5606.M34   1989
694—dc20   89-6564
                                                    CIP

FIGURE CREDITS

Figures 1-1, 1-2, 1-4, 1-8b, 1-12, 1-13a, 1-14, 1-16, 1-19, 1-20, 1-21, 1-23 courtesy of Stanley Tools, a division of the Stanley Works.

Figures 4-1, 4-10, 4-15, 4-16, 4-30, 4-41, 4-49, 4-53A, 4-53B, 4-57, 4-58, 4-59, 4-60, 4-63, 4-74, 4-76, 4-80, 4-81, 6-1, 6-4, 6-5, 9-1, 9-2 reprinted with permission of the Association of General Contractors of America and the Oklahoma State Department of Vocational-Technical Education.

Figure 6-2 reprinted with permission of the Western Wood Products Association.

Figure 6-6 reprinted with permission of the American Plywood Association.

CONTENTS

Part One—
Tools of the Trade

Part Two—
Standard
Carpentry Procedures

Part Three—
Carpentry Fundamentals

Part Four—Appendices

PREFACE

The construction field is in a constant state of change as a result of the development of new building techniques, materials, and processes. To successfully incorporate the latest technology with sound construction principles, a carpenter must have solid basic skills and knowledge of a wide range of areas.

The Carpenter's Toolbox Manual is designed as a guide for the beginner and a "hands-on" reference for the seasoned tradesperson. The alphabetical format allows quick and easy access to information ranging from tool descriptions and uses to the more complex aspects of building a sound structure.

This book is divided into four main parts. Each part builds on the previous section in order to give the reader a complete view of the carpentry trade.

Part I provides information on topics that could be considered the foundation of the trade. This section describes basic hand tools, as well as power tools that are commonly used on the job site.

Part II contains information on how to remedy carpentry problems, including techniques for troubleshooting construction problems and methods of creating the various elements of a structure.

Part III, on fundamentals, reviews basic mathematical principles, formulas for estimating needed materials, problems in identifying and selecting materials, and the various components that make up the building plans or blueprints.

Part IV includes a number of appendixes covering job safety rules and procedures, building codes and regulations, charts and tables to aid in calculating and comparing materials and a listing of trade associations.

The first two parts of this book provide lists of procedures that are essential to the trade. The emphasis throughout is on careful, safe equipment usage and construction techniques.

Gary D. Meers

ACKNOWLEDGMENTS

I would like to thank a number of organizations and people who have helped make this book a reality. The Associated General Contractors of America, the American Plywood Association, and the Western Wood Products Association were generous in granting permission to reprint various sketches, charts, and tables. The inclusion of their materials helped to clarify and illustrate the goals of this book. Permission was also generously provided by Stanley Tools, a division of The Stanley Works, to reproduce photographs of their products to illustrate selected hand tools.

I would like to also thank In Plain English, Inc. for patience, understanding, and expertise in taking the book from rough manuscript to finished product. It was not an easy task, but a most appreciated one.

Special thanks are also extended to Robert N. Hoffman for his careful review of the material for this book and his helpful suggestions.

In addition, I would like to thank all of the carpenters with whom I have worked over the years. They all had a hand in developing my skills so that I would have an opportunity to complete a task like this.

PART ONE

TOOLS OF THE TRADE

1
Hand Tools

Hand tools are the foundation of the carpenter's trade. Carpenters select their tools based on quality, function, and personal preference. Their tools become an important part of their lives, helping them perform their work quickly and accurately.

Over the years, hand tools have changed little in basic design, but the development of new materials has improved their cutting, gripping, and accuracy. Specialty tools are being developed all of the time, and tool displays or equipment outlets provide the best indication of the most recently developed tools and how they might be used for a particular job.

Quality tools are a must. The old adage that you get what you pay for can be applied here. Quality tools not only last longer but also allow carpenters to do the job more quickly and accurately, and less time is spent in sharpening, repairing, and replacing tool parts.

Because there are many different tools, a detailed study of the selection, care, and use of all the hand tools that are available is not possible. The tools discussed in this chapter are those most commonly used by carpenters on most jobs.

CLAMPING TOOLS

On the construction site, there are a number of clamping tools that are especially useful when you are

- completing finish work,
- gluing parts together,
- installing metal fasteners, or
- clamping jigs of fixtures to machines for special setups.

C-CLAMPS

C-clamps are made in many different shapes and sizes. The size represents the tool's largest opening. The depth is measured to the

(Figure 1-1)   C-clamp

back of the clamp and usually ranges from 1 inch to about 4 inches. The shapes are deep throat, square throat, and round. The clamping needs of the job determine the size and shape of the clamp to be used.

C-clamps are simple to use. They are attached and adjusted by means of a hand screw with a movable ball-joint pad. (See Figure 1–1.)

Be sure to select the correct size clamp for the job. A clamp that is too small can't get a proper grip and may slip off the work. Using a clamp that is too large can result in too much pressure being applied, thereby marring or breaking the stock. As a guide, select a clamp with a frame that is not more than 2 inches larger than the stock and protective scrap wood. Scrap wood should always be placed between the ball-joint pad and the stock to protect the work's surface against marring.

Be sure that the ball-joint pad is flat on the surface to be clamped. If pressure is applied when the clamp pad is at an angle, the clamp could slip and damage the work surface or injure the worker.

SPRING CLAMPS

Spring clamps are best suited for fast-setting glue jobs that require the quick application of many clamps to a number of pieces that need to be held together repeatedly, as when checking for fit. Clamp lengths are from 4 inches to over 8 inches, with the jaw width

(Figure 1-2)   Spring clamp

ranging from ⅞ inch to 3 inches. (See Figure 1–2.)

Spring clamps should be selected based on the proper jaw size for the job. They are simple to use because they operate by spring rather than screw action.

Some spring clamps come with vinyl-covered jaws to protect the stock. If you use spring clamps that don't have vinyl-covered jaws, place a piece of scrap wood between the jaws and the stock.

Make sure the spring clamp jaws are placed securely to prevent them from slipping and marring the stock or, if the tension is too strong, from flying off and causing injury. Some of the larger spring clamps have very strong springs that require two hands to open.

WEB CLAMPS

Web clamps (also called *band clamps*) are canvas or nylon bands that can be wrapped around a work and tightened by a crank or ratchet. (See Figure 1–3.) The band lengths commonly range from 12 to 15 feet.

(Figure 1-3)   Web clamp

Web clamps are used to hold together irregularly shaped work or to bring together several joints simultaneously. Tighten the crank or ratchet slowly while making sure that all of the shapes and joints are pulled together properly. There should be no twists or wrinkles in the web, since these can mar the work surface.

The web should be checked for tears or holes. If these weaknesses are found, the clamp should not be used until the web is replaced.

If a weak web breaks under pressure, it can cause the work to fly in different directions and possibly damage the work and injure the worker.

SPECIALTY CLAMPS

Specialty clamps are those that are made to perform specific but limited clamping functions. Examples of these are edge clamps and miter clamps. Edge clamps are used when the work is too wide or too long to use a C-clamp or bar clamp. The miter clamp (also referred to as a corner clamp) holds mitered joints for gluing or nailing. (See Figure 1–4.)

(Figure 1-4)   Miter clamp

As needs arise, new clamps are designed and become available to the public. Look for new clamps that make the job easier and quicker.

DRILLING AND BORING TOOLS

BIT BRACE

The bit brace is a hand operated drill that has three characteristics that at various times make it the tool of choice.

1. It can be operated where there is no electricity.

2. It is almost noiseless.

3. It has a ratchet that allows it to be used in cramped spaces.

The bits that are used in the bit brace are called auger bits, and each is selected based upon the requirements of the job.

Auger bits commonly come in sizes ranging from ¼ inch to 1 inch. The size is stamped on the tang or shank. (See Figure 1–5.) Often the size is given as a whole number, which indicates the size in sixteenths of an inch. The most widely used auger bits are the double twist, also called the Jennings pattern, and the single twist. The double twist leaves a cleaner hole; the single twist is stiffer and better suited for drilling deep holes or wavy grain.

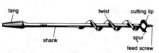

(Figure 1-5)   Common auger bit

After selecting the proper bit, secure it in the brace by tightening the jaws of the brace around the tang. (See Figure 1–6.) Locate and mark the center of the hole to be drilled. There are two ways to drill a hole through a board: (1) fasten a piece of scrap wood on the back, or (2) drill through the wood until the feed screw or tip of the bit emerges from the other side. Then back the bit out of the hole and continue drilling from the opposite side. This will prevent the wood from splitting when the full width of the bit is driven through the hole. (See Figure 1–7.)

(Figure 1-6)   Fitting a bit into the bit brace

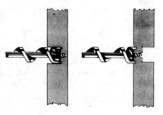

Be sure that the brace and bit are square with the hole. Placing a try square next to the bit will ensure a straight and accurate hole.

(Figure 1-7)   Boring a hole

SPECIALTY BITS

Countersink Bit. The countersink bit is used to create recess holes for flathead screws.

Expansive Bit. This bit is used for drilling holes ranging in size from 1 inch to 3 inches. A large screw located on the face of the bit is loosened and then the flange cutter is adjusted to the size of the hole that is desired. A hole gauge is located near the screw for easy adjustment. When drilling holes deeper than the bit length, use a bit extension.

Forstner Bit. This bit is used to make shallow, flat-bottomed holes. The bit must be carefully centered and tapped lightly into the surface before drilling the hole.

Screwdriver Bit. This bit is used to drive roundhead as well as flathead screws. A Phillips-style bit is also available. The screwdriver bit is also very handy for removing stubborn screws.

Spade Bit. A spade bit serves the same function as an auger bit but is much less expensive because it lacks the corkscrew track on the shank that helps carry sawdust out of the hole. A spade bit has a flat working surface (like a spade, hence the name) with a cutting edge on either side of the working surface. It is easy to use on wood since it has a sharp centering point that helps in accurately locating the center point of the hole to be drilled. Bit sizes range from ¼ inch to 1 inch in ¹⁄₁₆-inch increments.

FACTORS TO CONSIDER

- Always be aware of the pressure that is exerted against the work while drilling. Many soft woods will split or break if too much pressure is applied.

- Know where the bit tip is while drilling so that the wood can be reversed for final drilling before it is too late to prevent splitting.

- If a number of holes are to be drilled to the same depth, place tape around the bit and then mark the tape with a pencil to show the depth you want. If you don't mark the tape, you might forget whether you had intended to drill to the bottom or the top of the tape and end up with different hole depths.

- Always keep the brace and bit square with the hole to ensure accurate holes.

HAND DRILL

The hand drill is a crank-operated tool. A large wheel gear turns a smaller pinion, which then turns the drill chuck. It is used where there is no electricity on the site and/or where very light delicate drill work is required, for example, drilling pilot holes in thin narrow hardwood trim that is already in place. The weight of a power dril might cause the bit to angle and split the trim. The size of a hand drill is determined by the capacity of its chuck. Usually the chuck capacity is ¼ inch or ⅜ inch. Regular twist drills are used in the hand drill. A practical set of twist drills should include sizes of 1/16 to ¼ inch with increments (or jumps) of 1/32 inch.

Before drilling, use an awl to make a starting hole in the wood. This will keep the drill from wandering away from the spot where the hole is to be drilled.

The process of drilling the hole is very simple. Keep the twist drill square with the hole to alleviate side pressure. Don't exert too much pressure while drilling or the drill bit might break. Also ease up on the pressure just as the drill goes through the stock to prevent splitting.

FASTENERS AND DRIVERS

HAMMERS

Hammers come in a variety of shapes and weights and each is suited to perform a specific set of tasks. It's important to be familiar with all the types available, because the hammer is the most frequently used fastening tool. (See Figure 1–8.)

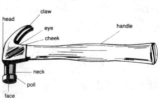

(Figure 1-8a)   Parts of a hammer

The two most common hammer shapes are the curved claw and straight (ripping) claw. The curved claw is best suited for pulling nails. The straight claw is designed for rough work and can be driven between boards and then used to pry them apart.

A hammer head is forged of high quality steel and is heat-treated to give the poll and face extra hardness. A bell-shaped, slightly

(Figure 1-8b)   Ball peen hammer

convex face minimizes hammer marks when nails are driven flush.

The size of a claw hammer is determined by the weight of its head. Claw hammer sizes range from 7 to 20 ounces. The 13-ounce size is popular for general-purpose work. A 16- or 20-ounce hammer is used for rough framing.

To use a hammer properly, grasp it near the end of the handle; this is the proper position for driving a nail. Hold the nail between the thumb and forefinger of the hand opposite. Place the nail on the work surface, and tap it lightly until it stands up firmly in the workpiece. Move your fingers away from the nail, and drive the

(Figure 1-9)   Using a hammer to pull a nail

nail in with smooth, even strokes. It is critical to be accurate because if the hammer head misses the nail, it will hit and mar the stock.

When using the claw of the hammer to pull nails, place a piece of wood under the head to avoid marring the workpiece. (See Figure 1–9.) Pry gently. If you exert too much

pressure the nail can fly from the hole with such force as to cause injury or damage. To pull large nails, use a wrecking bar or crowbar.

SPECIALTY HAMMERS

Hatchets. Hatchets are used for rough work on such jobs as making concrete forms or grade stakes. Some are designed to be used for roofing and shake-shingle work.

Mallets. These drivers have wood, plastic, rubber, or leather heads and are used to drive chisels and shape metal without marring the work surface.

Sledge Hammers. The heads on sledge hammers range in weight from 2 to 20 pounds and are used for heavy work, such as breaking up concrete sections.

FACTORS TO CONSIDER

- A quality hammer will last a lifetime. Since the hammer is the most commonly used carpentry tool, selecting the right one is critical. You must like its feel, balance, and grip before you make it a part of your tool kit.

- Do not use a hammer for work that it was not intended to do. For example, do not use a nail hammer for metal work.

- Strike only with the face of a hammer and don't use a hammer to hit anything that is harder than the hammer's face. The reason for this is that the impact power (the point at which the hammer face strikes another surface) can reach in excess of 300 pounds. This pressure can cause the hammer head to break or shatter.

- Use caution when pulling nails because the handle-leverage forces may reach several thousand pounds of pressure, which may cause the handle to break. If you must exert a lot of pressure, use a wrecking bar rather than risking injury to yourself or damage to your work.

- If your hammer has a wooden handle, check the head to make sure it is not loose. Excessive humidity or dryness can cause the head to loosen. If the head is loose, it can be tightened by driving a wedge (metal or wood) into the head opening and down into the handle. Fiberglass and metal handles need to be checked only occasionally since they are not subject to temperature or humidity changes.

- Always keep your hammer free of rust, since rust can damage the driving surface. Cover the hammer head with a light coat of oil to control this problem.

NAILSETS

Nailsets are used to conceal nails by driving finishing nails below the work surface. They are used in all phases of finish work, including both indoor and outdoor trim.

Nailsets are available in sizes from $1/32$ through $5/32$ inch in $1/32$-

inch increments and are either round-headed or square-headed. A square-headed nailset is easier to work with, since it won't roll away when placed on a sloping surface.

The nail that is to be set should be driven to slightly above the surface to prevent marring. Select a nailset that is slightly smaller than the nailhead, and drive the nailhead to a depth equal to its diameter. (See Figure 1–10.) Use putty or wood filler to fill in the recess that is left above the nailhead.

(Figure 1-10)   Using a nailset

SCREWDRIVERS

Screwdrivers are used to join workpieces or attach hardware to them. Screwdrivers come in a variety of sizes and tips. The sizes are specified by giving the length of the blade, measuring from tip to the ferrule (the metal band around the wooden handle at the point where the blade enters the handle) or, if the handle is made of plastic or metal, to the point at which the blade enters the handle. The 3-, 4-, 6-, and 8-inch screwdrivers are the most commonly used. The size of a Phillips screwdriver is given as a point number ranging from 0, the smallest, to 4. Size numbers 1, 2, and 3 fit most of the screws used in carpentry.

The slot of the screw and the blade of the screwdriver must be carefully matched. The width of the screwdriver tip should be equal to the length of the bottom of the screw slot. (See Figure 1–11.)

To drive small screws, make a starting hole with an awl. To drive larger screws and screws into hardwood, drill a pilot hole. The pilot hole should be slightly smaller than the shank diameter.

Once the screw is started, continue turning it while making sure that the screwdriver blade is seated squarely.

OFFSET SCREWDRIVER

This tool provides more leverage than straight-handled screwdrivers and can be used in tight or hard-to-reach places.

SCREW-HOLDING SCREWDRIVER

A screw-holding screwdriver is handy when working in hard-to-reach places. The screw is held in place by a clip or magnet while the screw hole is started.

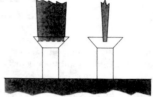

(Figure 1-11) Proper seating of a screwdriver

SPIRAL RATCHET SCREWDRIVER

A spiral ratchet screwdriver can drive a number of screws rapidly. Simply push down on the handle while the spiral ratchet spins the blade. Spiral ratchet screwdrivers come with various bits that can be mounted in the chuck to work with different screw sizes and types. They can also be set in reverse, to remove screws, or locked for use as an ordinary screwdriver.

FACTORS TO CONSIDER

- Always select the proper size screwdriver for the job. Screws that are driven with an undersized screwdriver are likely to suffer slot and head damage. An oversized screwdriver may fit the screw slot, but it might damage the work when used with a flathead screw.

- A screw follows the hole in which it is inserted, so the pilot hole must be straight and square.

- Keep the screwdriver properly seated in the screw head or slot at all times. Failure to do so might result in the screwdriver slipping and causing damage or personal injury.

- Do not use screwdrivers for anything except driving or removing screws. When screwdrivers are used as pry bars or wood chisels, they can be damaged and may become dangerous. A twisted or bent screwdriver is not only unsafe but also very hard to use and can damage materials and cause delays.

FILES AND RASPS

Files are a part of a carpenter's tool kit for two reasons: (1) they are used to keep woodworking tools sharp, and (2) they are used to shape and smooth wood surfaces.

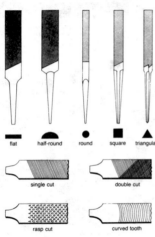

flat   half-round   round   square   triangular

single cut

double cut

rasp cut

curved tooth

(Figure 1-12)   Common files

Files are classified by shape and length and by the cut and coarseness of their teeth. Commonly used file shapes are flat, half-round, round, square, and triangular. The common cuts of teeth are single, double, rasp, and curved. (See Figure 1–13.) Additionally, special-purpose files are available. These are generally named for their function. The chain saw file, for example, is used to sharpen the teeth on the chain of a portable chain saw.

Files that are made in the United States are classified according to the coarseness of teeth (starting with the coarsest): coarse, bastard, second cut, smooth cut, and dead smooth. The coarseness is also affected by the length of the file, which is measured from the heel to the point. As the length increases, the tooth size increases. Therefore, two files that have the same coarseness designation but are different length will give slightly different finishes.

There are two main ways of holding a file during its use. The most common is to grasp the handle at the heel end of the file with one hand and the point between the thumb and forefinger of the other. The file is then placed against the surface to be shaped at an angle of approximately 20 degrees and moved back and forth along its own length, in a sawing motion. This method is used to remove material quickly or to achieve a rough finish.

For a smoother, finer finish you can draw the file. This is done by holding the file at both ends and placing it against the surface to be worked at a right angle. The file is then drawn back and forth along the surface without employing much lateral motion.

Any piece to be filed must be securely fastened in a clamp or vise. The work should be held at approximately elbow height, to facilitate the job.

FACTORS TO CONSIDER

- Never use a file that doesn't have a handle. This is to prevent injury to your hand during the filing operation.

- To produce a smooth surface, use a single-cut file and apply light pressure. To remove material from a surface quickly, use a double-cut file and apply heavier pressure. For fast removal, use a rasp and apply pressure appropriate to the hardness of the wood.

- There are several dozen types of specialty files available. To find out the types available and their uses, contact your tool supplier.

- Files should be cleaned with a file card and brush. For files used solely on metal, apply a coating of chalk dust to prevent metal particles from sticking to the file.

MEASURING AND LAYOUT TOOLS

Measuring and laying out a job properly ensure a successful start on a project. To perform these functions accurately, you must have the proper tools.

CHALK LINES

Chalk lines are used to mark long, straight lines, such as in subflooring layout. The end of the chalk-covered line is attached to a nail and held tight and close to the surface. The middle of the line is then lifted up and snapped. This action drives the chalk onto the work surface, leaving a distinct mark. Each time the line is reeled back into the case, it is automatically rechalked.

Since humidity can cause the chalk in the case to clump, always tap the case against an object prior to letting out the line to ensure that it is adequately coated.

Snap the line only once when chalking. If the line is snapped repeatedly, several lines will appear or the main line will be distorted.

LEVELS

Levels come in a variety of lengths and are made of a number of materials, based on their use. The body of the level may be made of wood, aluminum, or a special lightweight alloy. The length may vary from 3 inches (for a line level) to 6 feet. The standard level is 24 inches long.

A level is used by placing it against the work to be checked. The slightly curved vials in the level are filled with a nonfreezable liquid, and each vial has a bubble floating in it. The bubble moves within the liquid and rests at the level of the stock being checked. To check for proper placement of the stock, the stock is moved until the bubble in the level rests directly between the two marks on the vial.

The vials that run parallel to the long side of the level are called level tubes. These tubes are used for checking horizontal surfaces. The vials that run perpendicular to the length of the level are called plumb tubes, and are used to check vertical alignment. The vials on some levels are adjustable and are calibrated from 0 to 90 degrees so that they can be used to check pitched surfaces. Other levels have 45-degree tubes as well as level and plumb tubes.

(Figure 1-13a)   Carpenter's level

CARPENTER'S LEVEL

The carpenter's level is about 2 feet long and can be viewed from the top and side. It has bubble tubes for checking both vertical and horizontal surfaces. (See Figure 1–13a.)

ELECTRONIC LEVEL

An electronic level operates in the same manner as the standard level, except that it has LED readouts instead of vials to indicate plumb and level surfaces.

LINE LEVEL

The line level is 3 inches long and lightweight. It has hooks on either end so that it can be hung from a line for long-span leveling in grading and foundation work. (See Figure 1–13b.)

(Figure 1-13b) Line level

MASON'S LEVEL

This level is about 3 feet long, can be viewed from the side, and is made of metal alloy.

TORPEDO LEVEL

The torpedo level is small and can be used where other levels won't fit. Usually, it has a top-view, a side-view, and a 45 degree vial. (See Figure 1–13c.)

(Figure 1-13c) Torpedo level

FACTORS TO CONSIDER

- Always check the work surface to make sure that small objects, nails, or hardware do not hold the level away from the stock, causing the reading to be off.

- Avoid dropping or hitting a level. Although they are sturdy, well-made tools, they can stand only so much abuse.

MARKING TOOLS

CARPENTER'S PENCIL

A carpenter's pencil is oblong, with a flat, thick lead core. It is used to mark stock. The point is shaped by hand, and thus the carpenter can create a wide or narrow point as desired. Because the lead is thick, the point will not break as easily as that of a standard pencil.

MARKING GAUGE

The marking gauge is used to make a scratch along the length of a board so that the board can be cut to width. (See Figure 1–14.)

(Figure 1-14)   Marking gauge

TRAMMEL POINTS

Trammel points are used in pairs to mark large, circular layouts. Both points can be clamped to a straightedge. The trammel used for center must have a metal point, while the marking trammel can have a metal point or a pencil adapter for scoring or marking the work surface.

WING DIVIDER

The wing divider is used to mark smaller circular work. It has two metal points, and the scratching point may have a pencil adapter for marking. (See Figure 1–15.) A wing divider is also used to scribe a work-piece so it can be fit to an irregular shape. For example, if the side edge of a countertop must fit snugly against an irregularly shaped wall, the countertop should be marked and measured to fit by using a wing divider, compass, or scribe.

(Figure 1-15)   Wing divider

FACTORS TO CONSIDER

- Markers with metal points score the wood rather than mark it, so accuracy and setting must be done carefully.

- All pointed marking devices need to be stored and protected. Bent, damaged, or burred points can cause the marking efforts to be off or the device to jump out of the planned marking area.

PLUMB BOBS

A plumb bob is a metal weight that attaches to a suspended line. The weight of the plumb pulls the line into a true vertical position for layout. The point of the plumb bob always hangs directly below the point from which it is suspended. When the plumb bob is first suspended, it will rotate; when it stops moving, it is ready for checking.

SQUARES

There are a number of squares that must be used in laying out work. The selection of the appropriate square depends on the size of the work and the nature of the job.

COMBINATION SQUARE

The combination square is used to check the squareness of surfaces and edges and to lay out miter joints. The adjustable sliding blade allows it to be conveniently used as a gauging tool. (See Figure 1–16.)

(Figure 1-16)   Combination square

FRAMING SQUARE

This square is also called the *rafter square*. It is made of steel, aluminum, or steel with a copper or blued finish. It is both a marking square and a carpenter's calculator, since it is scaled so you can mark for rafter cuts according to roof pitch. The use of the framing square is illustrated throughout this book, especially in the sections dealing with rafter and stair layout.

T-BEVEL

The T-bevel has an adjustable blade that makes it possible to transfer an angle from one place to another. It is useful in laying out cuts for hip and valley rafters.

TRY SQUARE

Try squares are available with blades 6 to 12 inches long. Handles are made of wood or metal. These are used to check the squareness of surfaces and edges and to mark for right angle cuts.

FACTORS TO CONSIDER

- All squares are simple to use as long as they are firmly seated against the stock that is being marked.
- Care must be taken to keep the handle and blade square with each other.
- Squares should not be thrown or dropped, since this can knock them out of square.

STRAIGHTEDGES

A straightedge may be made of laminated wood or of metal. It is often used as an extension of the carpenter's level to check work that is longer than the level itself, such as walls or door frames. The straightedge is laid along the level so that it is in contact with the level's entire length before it extends beyond the level.

The straightedge can also be used as a guide in laying out plywood panels or in other work that requires a solid edge for marking.

Since the straightedge can be easily twisted, causing it to lose its accuracy, it is important to check that the edge is straight and square.

TAPES AND RULES

A badly measured project cannot be successfully completed. To ensure accuracy, the tape or rule must be read correctly, and the tools must be kept clean for easy reading.

FOLDING WOOD RULE

This rule is essential and is generally carried at all times while on the job. A standard size rule is 6 feet long. Those equipped with metal slides are useful in making inside measurements. Folding wood rules are used for general measuring and are especially handy where rigidity is needed, as in measuring wide window openings.

STEEL TAPE RULE

The steel tape rule ranges in length from 6 to 25 feet and is used when laying out larger projects. Because the tape is flexible, it can be used to measure round as well as straight objects. It is compact and can be clipped to work clothes and carried on the job. Long tapes are available in lengths of 50 feet and longer.

The metal tip of the steel tape rule slides back and forth to accommodate its own thickness, making up the difference for both the inside and outside measuring. This tip should be checked periodically to ensure that it has not become worn and slipped more than its width. If it has, either secure the tip in place or allow for the slippage in measuring.

FACTORS TO CONSIDER

- All rules must be placed on the workpiece carefully to ensure accurate measurements.
- Folding wood rules can break if folded incorrectly.
- Steel tape rules can sag or bend when taking long measurements. These variations can change the measurement by as much as a half an inch.

PLANING, SMOOTHING, AND SHAPING TOOLS

CORNERING TOOLS

Cornering tools are used to remove sharp corners from exposed wooden parts. Each end has a different sized opening, ranging from $1/16$ to $3/8$ inch.

PLANES

Planes are used to trim wood to size, to smooth it, to straighten irregular edges, to bevel and chamfer it, and, in special forms, to groove and shape it into moldings. The types of planes vary according to use. Many power tools are available that can do the job of hand planes much more quickly. However, certain planes are still very much a part of a carpenter's tool kit.

BLOCK PLANE

The block plane is the smallest shop plane (about 6 inches long). This plane is the one used most often by the carpenter and can be used with one hand. The blade or plane iron is mounted at a low angle, and the bevel of the cutter is turned up. The plane produces a fine smooth cut making it suitable for fitting and trimming work.

CABINETMAKER'S PLANE

This plane is also called a trimming plane and is $3\text{-}1/2$ inches long with an inch-wide blade. This plane is used for small and delicate work, such as removing excess wood from cabinet doors or drawer sides.

JACK PLANE

The jack plane is approximately 14 inches long with a 2-inch-wide blade. It is a general-purpose plane.

SURFORM PLANE

The Surform (a brand name) is a multiblade forming tool, that can be used as a plane or file for wood, aluminum, copper, brass, plastic, laminates, or other similar materials. Because it can be used to cut and smooth a variety of surfaces, the surface produced is rough and requires sanding to remove the tool marks. Surform tools come in three basic forms: the rasp, plane, and block type.

Basic Procedures for Using Planes

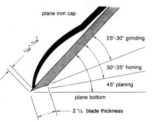

1. *When using planes that have adjustable blades, set the plane iron ¹⁄₁₆ inch back from the cutting edge. The plane iron cap must be set tightly against the cutting edge to prevent chips from entering between the cap and blade. The angle of the plane iron is adjusted for different types of jobs. (See Figure 1–17.)*

(Figure 1-17) Adjusting the cutting angle of a plane

2. *To cut an even width, align the blade along the bottom of the plane. Set the cutting edge parallel to the plane mouth by moving the blade on the block plane by hand or by using the lateral adjusting lever on a larger plane.*

3. *Start the plane flat on the work. Apply even pressure during the stroke to produce a continuous shaving. Hold the plane nose up at the end of the stroke to prevent a down curve.*

4. *Always plane in the direction of the wood grain to minimize chipping. Planing against the grain will leave a rough surface.*

UTILITY KNIFE

The utility knife is a multipurpose tool that can be used for shaping or shaving wood. The knife has interchangeable, retractable blades designed for various uses. Generally, the utility knife is used to shave down stock for a fit where smoothness is not the most important factor. There are no gauges on the utility knife, so the accuracy of work depends on the skill of the worker.

FACTORS TO CONSIDER

- Always keep the plane iron or blade sharp for cutting ease and minimum chipping.

- To hone a plane iron, use an oilstone (or whetstone). Hold the blade at about a 30-degree angle to the stone, rubbing the blade back and forth along the stone's length.

- Make sure that the wood you are working on is free of nails or other metal. Hitting a nail can take a chunk out of the plane iron, which will then have to be reground or thrown away.

- When planing across the end grain of a board, set the plane iron for a shallow cut. It's important to first bevel the ends of the board slightly. If you don't, the plane will split the wood as it nears the end of the cut. Make the bevel cut with the plane at a 45-degree angle.

- Another way to prevent splitting of end grain corners is to tightly clamp a piece of scrap wood on each side of the workpiece so that it's flush with the surface being planned. By doing this, the scrap, rather than the work, will split.

- Different woods have different grain patterns and textures. Be familiar with these characteristics, so you know what to expect when you plane the various types of wood.

WOOD CHISELS

Wood chisels are used to trim and cut away wood or composition materials to form joints or recesses. They are also useful in paring and smoothing small interior surfaces that cannot be reached by other edge tools. Width sizes range from 1/8 to 2 inches. The sizes most frequently used are 3/8, 1/2, 3/4, and 1 1/4 inch. Most chisels have plastic handles that withstand the strike of a hammer or mallet to make deep cuts. (See Figure 1–18.)

BUTT CHISEL

The butt chisel is the shortest chisel (about 7 to 9 inches long) and is used in tight spaces.

FIRMER CHISEL

The firmer chisel has a thick blade for heavy driving.

GOUGE

The gouge has a hollow blade and is used to create grooves for fitting work or for decoration.

(Figure 1-18) Two methods of using a chisel

MILL CHISEL

The mill chisel is the largest chisel (16 inches long) and is used for heavy-duty work in rough construction.

PACKET CHISEL

The packet chisel is 9 to 10½ inches long and is used for general shop work.

FACTORS TO CONSIDER

- Never work with a dull chisel.
- Sharpen chisels on an oilstone at a 30-degree angle.
- Maintain control of the chisel at all times, and do not let the edge of the chisel dip down while striking the handle. Dipping the edge of the chisel may cut the wood too deeply or split the wood.
- Carefully store and carry chisels to prevent their sharp edges from causing damage or injury.

PARING CHISEL

The paring chisel has a thin blade with a 25-degree edge. It is driven by hand and used to make precise shaping cuts for fitting work.

PLIERS AND WRENCHES

Your tool kit would not be complete without one or two pairs of pliers and several wrenches. Pliers and wrenches come in a variety of sizes and shapes, and choosing the right tool depends on the nature of the job.

PLIERS

Pliers extend the worker's grip and hand length. The shape, size, and function of pliers are based on the task at hand, as well as the inventiveness of the worker. Periodically new pliers are introduced in the market because someone has found a new or faster way of using these gripping devices.

CHANNEL-TYPE PLIERS

Channel-type pliers have a multiposition pivot that permits the jaws to be adjusted for gripping objects that are up to 2 inches wide. These pliers can grip any shape.

LOCKING PLIERS

Locking pliers grab and hold work when a lever on the handle is engaged. This grip is maintained until the lever is disengaged. These pliers come in various sizes and shapes and are designed to meet different clamping needs. They are especially useful when a strong grip is needed to hold an object for nailing or welding.

SLIP-JOINT PLIERS

Slip-joint pliers are so named because of the two-position pivot which allows them to be used with normal or wide-jaw opening. They are useful for general gripping work. (See Figure 1–19.)

(Figure 1-19) Slip-joint pliers

WRENCHES

Wrenches are used to grip nuts and bolt heads that must be held for fastening and unfastening. They are available in a variety of forms and are designed to grip everything from pipes to spark plugs. They are also used for work in hard-to-reach places.

When selecting a wrench for a particular job, make sure that it is the correct size. A wrench may appear to be the proper size because it is wedged on the points of the nut. When pressure is applied to a wrench that is the wrong size, the wrench will slip and ruin the nut.

You should include a variety of wrenches in your tool box. You'll save time and wear-and-tear on materials by having the right wrench available when needed.

ADJUSTABLE WRENCH

The adjustable wrench has a movable jaw that can be set by turning an auger screw. It is an all-purpose wrench used to adjust equipment or to secure a variety of bolts.

Using an adjustable wrench in these situations saves the time needed to change to a new box or open-end wrench for each adjustment. The size of the

(Figure 1-20) Adjustable wrench

wrench is determined by the length of the handle, which in turn determines the width of the jaw opening. The smallest handle length is 4 inches; the largest 24. (See Figure 1–20.)

ALLEN WRENCH

An Allen wrench fits hexagonal recesses in various hex socket head screws.

NUT DRIVER

A nut driver has a socket that can grip nuts on one end and a screwdriver type of handle on the other. Some may have interchangeable sockets. They come in sizes from $\frac{3}{16}$ to $\frac{1}{2}$ inch.

PIPE WRENCH (STILLSON WRENCH)

A pipe wrench is used on round objects such as pipe. It has a movable jaw that tightens as pressure is applied to the handle.

SAWS

Saws come in a variety of sizes and shapes. Each performs a specific function. The function of all saws is to separate one part of the material from another.

BACKSAW

The backsaw is designed for joint-cutting work. It is 10 to 16 inches long and usually has 12 or 13 teeth per inch. It provides a smooth cut when used to cut with or across the grain of the wood. The miter box saw is a longer (up to 26 inches) version of the backsaw that has 11 teeth per inch. (See Figure 1–21.)

(Figure 1-21)   Back saw

When using the backsaw for cutting outside the miter box, always start with a backward cut after securing the work. When using the backsaw in a miter box, first mark the work for the cut and then line up the mark with the slots. Be sure and cut on the waste side of the line. Hold the work against the back of the box and start with a back stroke holding the handle end tilted slightly upward. Level the saw as you proceed.

FACTORS TO CONSIDER

- Keep the saw at a right angle or vertical to the work for a square cut.
- Never start the cut by pushing forward because the back side of the work may split.

CROSSCUT SAW

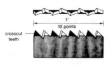

Every carpenter needs a good crosscut saw that has a tooth size of 8 to 11 points. Tooth sizes in saws are given as points, or teeth per inch. (See Figure 1–22.) Crosscut saws, as the name implies, are designed to cut across the wood grain. Their teeth are pointed and have the same effect as a knife cutting into the wood fibers. In quality saws, the teeth are precision-ground to tiny points. In low quality saws, the teeth have the same shape as those better-grade saws but are not precision-ground for clear cutting. If a saw has a low point number, it cuts fast but leaves a rough surface. A saw with a high point number works more slowly but leaves a smoother surface. To reduce friction and boost efficiency, alternate teeth are set (bent) outward about a quarter of the blade thickness to opposite sides. This produces a cut that is slightly wider than the blade thickness and so lets the saw cut freely.

(Figure 1-22)   Crosscut saw teeth

Basic Procedures for Using a Crosscut Saw

1. *Lay out and mark the stock to be cut.*
2. *To begin the cut, place the butt portion of the blade near the handle on the waste side of the line on the board to be cut. Use several pulling strokes to make a starting groove or kerf.*

3. *Don't cut on the marked line, cut on the waste side.*

4. *Continue with full strokes for fast cutting and even distribution of tooth wear along the blade.*

5. *The saw should be held at a 45-degree angle from the board's edge.*

6. *Since the crosscut saw cuts on both the forward and back strokes under its own weight, you need apply only light pressure in using it.*

7. *When completing the cutting of stock, support the waste piece while finishing the cut with short light strokes to prevent splitting.*

FACTORS TO CONSIDER

- When not using a saw, store it in such a way as to protect its teeth from wear or damage.

- Cut plywood with a crosscut saw, regardless of the direction of the surface grain.

- Let the saw glide over the work. You are using a saw that has knife-like points, so don't try to force the saw through the wood—let it cut its way through the wood.

DRYWALL OR SHEETROCK SAW

This saw has large, specially designed teeth for cutting through the paper facings, backings, and gypsum core of

(Figure 1-23)   Drywall saw

drywall. The saw has rounded gullets (openings in the blade) to prevent its clogging. (See Figure 1–23.)

Mark the line you wish to cut, then place the point of the saw blade where you wish to begin. Slowly work the blade up and down or left to right, depending on the direction you intend to cut. Apply more pressure as you force the blade tip through the drywall. However, don't apply too much pressure, since this might break the drywall or cause damage behind it. After the blade tip has passed through the drywall, start the sawing action while following the marked line.

HACKSAW

Hacksaws are used to cut nails, bolts, and other metal fasteners, as well as trim used on both exterior and interior work. Most hacksaws have adjustable frames to fit several sizes of blades.

Hacksaw blades are made of high speed steel, tungsten alloy steel, molybdenum steel, and other special alloys. Generally, blades with coarse teeth are used on thick metal, and blades with finer teeth are used on thinner metal. At least two teeth should always be in contact with the material to prevent a thin section of metal from hooking between the teeth and breaking them. When cutting very thin stock, angle the saw so that the teeth are in contact with a part of the surface, rather than the edge.

As a rule, you should use a blade with 14 teeth per inch for brass, aluminum, cast iron, or soft iron. A blade with 18 teeth per inch is recommended for mild steel, tool steel, or general work.

Like wood-cutting saw blades, hacksaw blades have a set or angle in the direction of the teeth. This provides clearance for the blade to slide through the cut. Blades should be installed with teeth pointing forward, away from the handle.

Basic Procedures for Using Hacksaws

1. *Fasten the work in a vise or jig and make sure it is secure.*
2. *Grip the handle of the saw in your right hand and hold the front of the frame with your left.*
3. *Use moderate downward pressure on the forward cutting stroke and almost no pressure on the return noncutting stroke.*
4. *Use as much of the blade length as you can for maximum speed and minimum tooth wear.*
5. *Saw steadily at about 40 to 50 strokes per minute.*

KEYHOLE AND COMPASS SAWS

The keyhole saw derives its name from its original use—cutting keyholes. A companion to the keyhole saw is the compass saw, which was also used to cut keyholes. Today, both saws have a variety of uses. The typical compass saw has 12- to 14-inch blades with 8 to 10 teeth per inch. Keyhole saws have narrower blades, usually 10 or 12 inches long, with 10 teeth per inch. Either saw can cut curves, though keyhole saws can cut smaller circles than compass saws. Because keyhole and compass saws do not have outside frames to hold blades, they are not limited, as a coping saw is, to working near the edge of a panel. Therefore, they can be used to cut openings in floors or walls for pipes or electrical outlets, with the cut starting from a bored hole. (See Figure 1–24.) A vertical stroke is used to begin the cut. As the cut progresses, the saw is brought to about a 45-degree angle. If you start a cut from the edge, the saw can be at that approximate angle from the beginning.

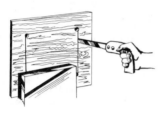

(Figure 1-24) Using a compass saw

RIPSAW

The ripsaw has chisel-shaped teeth that are designed to cut parallel to the grain. Usually, the saw has a 26-inch blade with 5-½ teeth per inch. (See Figure 1–25.) The teeth move the wood fibers out of the way in a chopping fashion, rather than cutting the fibers in the manner of a crosscut saw. Alternate teeth are set outward about a third of the blade thickness to widen the cut and reduce friction in the kerf. The ripsaw is no longer considered a "must" in a carpenter's tool kit, since most ripping operations are performed with power saws.

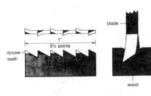

(Figure 1-25) Rip saw teeth

Basic Procedures for Using a Ripsaw

1. *To start the cut, use the tip portion of the blade (not the butt as with the crosscut saw). Quality ripsaw teeth are one point finer at the tip.*

2. *Begin with a few short, pulling strokes and then continue with full strokes.*

3. *Hold the saw so that the blade is at approximately a 60-degree angle from the work surface.*

4. *If on long cuts, the saw veers away from the line, flex it slightly toward the line to bring it back on course. Avoid sharp bending.*

5. *Finish the cutting operation with short, light strokes and then take full strokes at the end of the workpiece to prevent splitting.*

CARE AND MAINTENANCE OF HAND TOOLS

The pride carpenters take in their work is invariably reflected in the condition of their tools. Much money is invested in acquiring a comprehensive assortment of quality hand tools, and so an appropriate amount of time and effort should also be devoted to maintaining them. Well-kept tools not only reflect the work habits and skill of the owner; they also help in completing jobs faster, safer, and more accurately. Never use broken, dull, or damaged tools; they can make the job difficult to complete and can be dangerous as well.

CLEANING AND STORING

Wipe all tools clean after use. Occasionally, saturate the wiping cloth with a light oil, such as lemon oil furniture polish. This oil not only cleans tools, but also leaves a light film that protects the metal surfaces from rust.

Clean and tighten all tool handles. Replace those that are broken or split.

Keep all tools in compartments or holders that prevent them from hitting each other while being transported. This will not only protect their edges or teeth. It will also prevent their handles or other work surfaces from being marred or split.

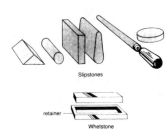

Slipstones

retainer

Whetstone

(Figure 1-26)   Honing instruments

HONING

Whetstones and slipstones are used to sharpen, or hone, tools. These stones come in a variety of shapes and sizes and should be selected based on the tools to be sharpened. (See Figure 1–26.)

A whetstone is rectangular and is stored in a wooden container. The best whetstone for general-purpose use is one with both a coarse side and a fine side. The whetstone should be lubricated with a light oil. The tool is then passed over the stone for honing. Loose grit and sludge can be removed from the stone with a solvent-soaked rag.

A slipstone is selected according to the shape of the tool that it will be used on. The slipstone is passed over the tool for honing.

Basic Procedures for Honing Hand Tools

1. *Hone edge tools on a whetstone. For single-bevel edge tools like chisels and planes, place the tool on the stone with the bevel flat on the surface. Raise the back edge of the tool a few degrees so only the cutting edge is in contact. Then move the tool back and forth until a fine wire edge can be detected by pulling a finger over the edge. Now place the back of the tool flat on the whetstone and stroke lightly several times. Turn the tool over and again stroke the beveled side lightly. Repeat this operation several times until the wire edge has disappeared from the cutting edge.*

   A cutting edge can be honed a number of times before grinding is required. When the bevel becomes blunt, reshape it with a grinding operation. The grinding angle for tools will vary somewhat, depending on the work the tools must perform. The angle for grinding plane irons is given in the section on planes. For chisels, the angle of grinding can be determined from another chisel that does not have a blunt bevel.

2. *A number of tools may be sharpened with a file. For auger bits, use a special auger-bit file. Sharpen the lips or cutters by stroking up*

through the throat. Do not file the underside. File the inside of the spurs, keeping them at the same length.

3. *Saws require filing as well as setting. Before filing, they often require jointing. In this operation, the height of the teeth are struck off evenly. Filing a saw is a tedious operation. Most carpenters prefer to send their saws to a shop where they can be machine sharpened.*

4. *Screwdriver tips should be kept square. When they become rounded, they can be filed straight across the blade until they are again square.*

2
Power Tools

Power tools, like hand tools, make up a very important part of a carpenter's tool box. They allow carpenters to do their work faster, easier, and, in many cases, more accurately. Power tools come in as many forms as there are tasks to be performed. This section deals with the most commonly used tools. Your tool supplier is a good source for information on the newest specialty tools and their uses.

There are two general categories of power tools: portable and stationary. Portable tools are lightweight, hand-held during operation, and easily carried. The tools are brought to the work. Stationary tools (or machines) are mounted on benches or stands that rest firmly on a work surface. The work is brought to the machine.

Safety is a primary concern when operating power tools. A later section in the book deals with general power tool safety, but, where appropriate, safety tips are given for specific tool usage. An overriding concern during the operation of any power tool is electrical safety. The potential for electrical shock is one of the major hazards of using power tools and steps must be taken to prevent it.

All tools must be properly grounded. Plugs and cords must be of approved types. Portable power tools should be double insulated or otherwise grounded to protect against shock. A ground fault circuit interrupter (GFCI) should be used on all construction sites. It can be installed in a circuit or can be plugged into an outlet that is grounded. The unit "senses" when a ground fault has occurred and will turn off power to the tool.

DRILLS

The portable electric drill has a wide variety of uses, and it is the most frequently used power tool. It can drill metal, wood, plastic, and concrete; drive and remove screws; polish surfaces; and perform many other functions.

Portable electric drills are designated by the maximum drill shank diameter that the chuck can accommodate. Sizes vary from ¼ to 1 inch. The drill speed decreases as the size of the drill chuck increases. A good all-purpose portable drill is one that has a ⅜-inch chuck, variable speed, and is reversible.

Variable-speed drills allow the operator to set the trigger switch to any desired speed via an adjustable knob. This feature allows the drill to be set for proper use and thus increases the options for using the drill on the construction site.

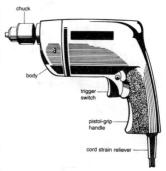

(Figure 2-1a)   Pistol grip drill

TYPES

The three basic types of portable drills are the pistol-grip drill, the D-handle drill, and the spade-handle drill. The pistol-grip drill is the most commonly used because it is designed for one-hand operation. Various models of this type are available with different speeds and chuck capacities. A ¼-inch or ⅜-inch chuck is generally selected by the carpenter. The average drill speed is rated at about 2,000 RPM; 1,000 RPM is best for woodworking. (See Figure 2–1a.)

The D-handle drill gets its name from the shape of its handle. It is designed for drilling small holes. Its advantage over the pistol-grip drill is its auxiliary handle, which permits more accurate drilling. (See Figure 2–1b.)

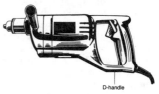

(Figure 2-1b)   D-handle drill

The spade-handle drill also is named for the shape of its handle.

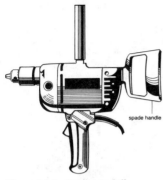

(Figure 2-1c)   Spade grip drill

It has the largest chuck capacity, taking up to a 1-inch drill bit. It operates at a slower speed than the other two types of drills because it is designed for heavy-duty use. Its average drill speed is rated at about 600 RPM without a load. Because of its size, both hands are required to use it. (See Figure 2–1c.)

Cordless portable drills, have been developed with rechargeable self-contained nickel-cadmium battery power packs for use on jobs where there are no power lines. If extensive use of the cordless portable drill is going to be made, it's advisable to have two battery packs. This will allow you to charge one power pack while the other is being used.

PARTS

Drills have three major parts that the operator must be familiar with: body, handle, and chuck. The body of a drill should be made of either lightweight metal alloy or high impact plastic. Either substance must be strong enough to resist the day-to-day abuse of construction work.

There are several handle shapes. All are designed to give the operator easy control of the drill. Optional handles may be added as needed for specific jobs.

Portable drills come with two kinds of chucks: key and keyless. The three-jaw chuck is designed to center the drill bit exactly. The jaws are opened by turning the sleeve counterclockwise (as seen from the bit end). After the bit is inserted in the chuck opening, the sleeve should be turned clockwise until the bit is gripped in the chuck jaws.

The chuck key should be inserted in one of the three holes of the chuck body and turned clockwise until the jaws are tight. It should then be fitted into each of the other two holes until the bit is secure.

A keyless chuck works in much the same way as the key chuck, except the bit is tightened firmly in the chuck by hand. This chuck works especially well when small drill bits are used.

The types of drill bits that work with the portable electric drill are:

1. straight-shank twist-drill bits for use with metal, plastic, and wood;

2. spade, or power-type, bits, which are used to bore holes in wood;

3. combination wood-drill and countersink bits, which make pilot and shank holes for screws and which countersink for screwheads in one drilling operation; and

4. countersink bits, which drill a countersink to accommodate flathead screws.

Basic Procedures for Using Portable Power Drills

1. *Select the correct size and type of bit. Fasten it in the chuck.*

2. *Connect the drill to a properly grounded outlet.*

3. *Mark the center point of the hole using an awl or nail. This will prevent the bit from wandering and marring the workpiece.*

4. *Put the point of the bit in the dent made in the workpiece. Do this before you turn on the drill.*

5. *Hold the drill perpendicular to the work. Setting a try square along the drill will help you align the drill. (See Figure 2–2.) After doing this a few times you will be able to feel how the drill should be held to keep the required alignment.*

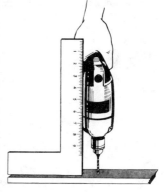

6. *Drill the hole. Use just enough* (Figure 2-2) Drilling angle
pressure to keep the drill cutting into the wood. Let the tool do the

(Figure 2-3)   Drilling using a back block

work, while you apply steady, even pressure.

7. *Use a block of wood to back the board if the hole is to be drilled completely through. (See Figure 2–3.)*

8. *Withdraw the bit from the hole with the motor still on and then turn off the switch.*

FACTORS TO CONSIDER

- When using a spade bit, either of two procedures will prevent splitting the stock: (1) use a back board or (2) drill until the point of the bit comes through the back of the board, then reverse the board, place the point of the bit in the hole and complete drilling.

- When drilling a hole completely through wood, make sure you know what is behind the workpiece. Drilling too deeply may put the bit through the back board and into a pipe, electrical line, or finished surface.

- Depth gauges are available that hook onto the shaft of the drill bit at any required distance. This enables you to drill multiple holes to the same depth.

- Never force the bit through material.

- The power drill can be used for many different purposes, such as polishing, sanding, grinding, and mixing. Check with your tool supplier for useful attachments such as flexible shafts for work in tight places, right angle attachments for use in limited work areas and grinding discs for paint and rust removal. Also available are drill holders that can secure the drill to the work area for use as a stationary tool for such operations as grinding.

SAFETY PROCEDURES FOR USING POWER DRILLS

Layout. Carefully plan and lay out your work before drilling holes.

Eye protection. Wear goggles when using power drills, regardless of the material you are working on.

Drill bits. Make sure you have selected the proper size and type of bit for the job. Do not use larger bit sizes than are recommended by the drill manufacturer.

Chuck. Fasten the bit securely in the jaws of the chuck before attempting to do any work.

Adjustments. Disconnect the plug from the power outlet before you make adjustments or change bits.

Electrical grounding. Check that the electrical connection is grounded.

Air vents. Keep the air cooling vents on the drill housing free of dirt and sawdust.

Material. Make sure the stock is held firmly so it will not move during the operation.

Drilling. Turn on the switch for a moment to see that the bit is properly centered and runs true. Hold the drill firmly in one or both hands and at the correct drilling angle. Keep the drill aligned with the direction of the hole to avoid forcing and breaking the bit.

Deep holes. When making deep holes, especially with a twist drill, withdraw the bit several times during the drilling operation to clear the cuttings.

Stopping the drill. Hold the drill until the switch is turned off and the motor has stopped. Then place it on a firm surface where it will not get knocked off.

JOINTERS

The jointer is an electrically driven power planer that edges and surfaces lumber by means of a three-bladed cutter head revolving between the tool's front and rear tables. The size of the jointer is determined by the width of the knives used in the cutter head and ranges in even numbers from 4 through 16 inches. Most carpenters select jointers 4 to 6 inches in size because they can be easily moved from one job to another and yet are large enough to do most of the work required.

In order to create a smooth even surface, the cutter head revolves between 3,600 and 4,000 RPM with the three knives in equal adjustment with each other. If one knife is set higher or lower than any of the others, the work surface will be nicked or uneven.

PARTS

The three main adjustable parts are the infeed table (front table); the outfeed table (rear table); and the fence.

The surface of the rear table must be level with the cutting height of the knives. If it is higher or lower, the planed edge or surface will not be straight. The front table is easily adjustable. It can be lowered to provide the depth of cut on the stock.

The fence is used as a guide. It is usually set at a 90-degree angle to the table to get edges planed at a right angle to the face. It can be set at an angle to produce a chamfer or a bevel.

The jointer has a spring action guard that covers the cutting knives. It swings out as the board is planed, thereby protecting the operator. This guard is to be in place at all times.

The jointer is most often used in constructing cabinets, built-in shelves, or other work where a smooth, finished, tight-fitting joint is desired.

Basic Procedures for Using Jointers

Planing a Surface

1. *Adjust the infeed table for a cut of about ¹/₁₆ inch.*

2. *Check to see that the guard is in place and working properly.*

3. *Check the surface of the board for a warp or slight twist. The concave face should be placed down and planed first.*

4. *Turn on the switch, and allow the machine to reach full speed.*

5. *Push the stock forward firmly with both hands. After about 12 inches has been planed, move your left hand forward slightly beyond the cutter head. Stand at the left of the front table as you plane. Plane in the direction of the grain. Use a push block for planing surfaces of a narrow or short board.*

6. *Check the planed surface to see that it has been fully planed and is smooth. If necessary, make additional cuts until the warp or rough-sawed finish is removed. The final cut should be very shallow.*

Jointing an Edge

1. *Check the fence with a try square to make certain it is set at a right angle to the table.*

2. *Select the best edge of the board to be planed. This should be the straightest one with the fewest irregularities.*

3. *Adjust the depth of the cut to approximately $1/16$ inch. This is done by lowering the front table.*

4. *Turn on the switch and allow the motor to reach full speed.*

5. *Place the board on the front table with the best surface (face side) against the fence. Make certain you plane with the grain.*

6. *Hold the board against the fence firmly with both hands, and slowly push it over the cutter head.*

Jointing an End

1. *Check the end of the board with a try or framing square.*

2. *Adjust the depth of the cut by raising or lowering the front table. The depth of the cut should be very shallow—approximately $1/32$ to $1/16$ inch. A deeper setting will tear the grain at the end of the cut.*

3. *Turn on the switch, and allow the motor to reach full speed.*

4. *Place the end of the board on the front table with the face side against the fence. Remember, never plane the end grain with a jointer if the board is less than 8 inches wide.*

5. *Hold the board firmly with both hands; slowly push it forward over the cutter head.*

6. *Another method of jointing an end is first to make a short cut of about 1 inch along one end. Reverse the board and joint the end to blend with the cut.*

SAFETY PROCEDURES FOR USING JOINTERS

Sharpness. Be sure that the jointer blades are sharp.

Adjustments. Never make any adjustments while the jointer is in operation. Check the jointer manual for instructions on adjustments.

Cut. The maximum cut for jointing on a small jointer is ⅛ inch for an edge and ¹⁄₁₆ inch for a flat surface.

Planing. Stock must be at least 12 inches long. Stock to be surfaced must be at least ⅜ inches thick unless a special feather board is used.

Direction of cut. Always try to plane in the direction of the grain.

Starting the motor. Always allow the jointer motor to reach full speed before starting to plane.

Hands. Keep your hands away from the cutter head even though the guard is in position. Maintain at least a 4-inch margin of safety.

Stopping the jointer. When work is completed, turn off the motor, and stand by until the cutter head has stopped.

Electrical grounding. Be sure that the electrical connection is properly grounded.

FACTORS TO CONSIDER

- Always make sure the knives are in equal adjustment with each other. Run a piece of scrap stock over the jointer and check for uneven surfaces. This check should be made each time the jointer is moved from one location to another. While being transported, the knives may have been knocked out of alignment.

- Different kinds of wood, grain pattern, and texture require that you move the stock over the knives at different speeds.

- Never run dirty, painted, or wet stock over the knives; running such stock will dull the knives quickly.

- Make sure all hardware is out of the stock that is being run. Hardware like nails or screws will dull the knives and may even ruin them by taking out large chunks of the cutting surface.

- Check all adjustments and guards. Make sure everything is in good working order before operating the jointer.

- There are many other operations that can be performed with the jointer such as planing chamfers, bevels, rabbets, tapers, and tenons.

PLANES

The portable power plane is actually a portable electric jointer. It has become an important labor-saving tool because of the speed and accuracy with which it can be operated.

TYPES

There are two forms of the power plane that are commonly used by the carpenter. These are the hand plane and the block plane. Both forms use a cutter composed of a solid body with two spiral cutting edges ground on them. This cutter body can be removed and sharpened with a special tool that can be purchased from the manufacturer of the plane. These planes range in size from the 3¾-pound block planes to the 16-pound heavy-duty hand planes. Cutter speeds range from 18,000 to 25,000 RPM.

The block plane is designed to do such operations as trimming the edges of doors, rabbeting cabinet doors, beveling countertop edges, planing shingles, and fitting drawers. The block plane is light enough to be operated with one hand. It can plane 2-inch dressed lumber with a maximum cut of ¹⁄₆₄ inch on each pass. A high speed steel cutter is used for all wood surfaces. With a carbide cutter, the plane can be used on building materials such as plastic laminates, hardboard, and aluminum.

The power hand plane does the work of the jack plane but in a quicker and less tiring fashion. It also produces a much smoother surface on knots and wavy grain. It is most commonly used for squaring and cutting down long edges such as door edges, molding, or trim.

The plane has a fence that projects downward from one side, perpendicular to the bottom. When you are planing the edge of a board or door, this fence is held against the side of the work to produce a perfectly square-edged cut. The fence can be removed for surface planing of broad areas, and also can be tilted for bevel planing.

PARTS

The basic parts of a power plane are described below.

Body. The body is the housing that encloses the motor and the spiral cutter. It is usually made of lightweight aluminum alloy.

Wrap around handle. Handles vary in size and shape, depending on the size and model of the portable plane. The handle, located for easy use of this tool, gives the operator control of the trigger switch.

Depth adjustment lever. The depth adjustment lever controls the depth of cut. The markings are calibrated to show the exact depth of cut.

Cutter blade. The cutter blade is a solid piece of specially hardened steel upon which are ground two spiral edges.

Chip deflector. This is the part of the body that throws chips out to the side.

Bevel adjustment lever. The bevel adjustment lever permits the plane bed or fence to be set to make outside bevel cuts from zero to 15 degrees and inside cuts from zero to 45 degrees. The adjustment is usually made by loosening a lever or wing nuts.

Basic Procedures for Using Power Planes

1. *The operating stance for running a power plane is similiar to that used with a nonpower plane.*
2. *The workpiece must be securely fastened before operation begins.*
3. *Adjust the depths for making the cut.*
4. *Plug the electric cord into a grounded power outlet.*
5. *Grasp the plane so that the operating hand has a solid grip on the handle and the forefinger is free to control the switch.*
6. *When using a block power plane, grasp the machine firmly with the operating hand and use the other hand to steady the workpiece.*
7. *Place the plane on the board with the cutter slightly back from the edge of the wood. Be sure that the electric cord cannot interfere with the planing process.*
8. *Turn on the switch. Push the plane to make the cut. Keep pressure on the front shoe with the nonoperating hand to ensure an accurate and even cut.*
9. *Continue planing. Maintain even pressure with both hands (for a hand plane) until you have almost completed the cut.*
10. *To complete the cut, keep greater pressure on the rear shoe than the front.*

11. *A final cut of* $1/32$ *inch will give a smooth surface that will probably not require sanding.*

12. *A bevel can be cut by adjusting the fence to the desired angle. The procedure for planing a bevel is the same as for straight planing.*

FACTORS TO CONSIDER

- Make sure the cutter is always sharp. Dull cutting edges can cause burns on the workpiece.

- As the plane is moved across the wood be aware of wavy grain or knots, since these can cause the plane to jump or move off square.

- Make sure all hardware, such as nails and screws, has been removed from the workpiece.

- Check all adjustments and guards. Make sure everything is in good working order before operating the plane.

- Study the operator's manual for details on the proper use and maintenance of your particular plane.

SAFETY PROCEDURES FOR USING POWER PLANES

Grounding. Be sure that the electrical connection is properly grounded.

Electric power. Check that the power switch on the plane is in its "off" position before you connect the plane to the power supply.

Stock. Always clamp the work securely in the position required to perform the operation.

Hands. Do not attempt to operate a power plane with one hand if the plane was designed to be operated with two.

Cutting depth. Always check for correct depth adjustment before making a cut.

Turning off power. Remember to turn off the switch before taking either hand off the plane after you have made a cut.

ROUTERS

The portable router is a simple power tool consisting of a high-speed electric motor mounted vertically on a horizontal bare plate.

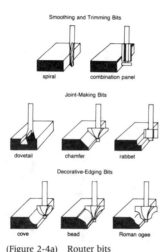

Smoothing and Trimming Bits

spiral   combination panel

Joint-Making Bits

dovetail   chamfer   rabbet

Decorative-Edging Bits

cove   bead   Roman ogee

(Figure 2-4a)   Router bits

The portable router cuts into and through wood and many other materials to a desired thickness and depth. Router accessories make it possible to produce intricate joints, decorative cuts, and inlays. The router can also be used to shape edges, cut recesses for door hinges, and make dovetail joints.

The wide selection of bits and cutters make the portable router an extremely versatile tool. It can be used for freehand cutting by guiding it with the hands alone, or it can be used with various templates.

The size of the portable router is measured by the horsepower (HP) rating of its enclosed motor. This varies from ¼ to 3 HP. The speed range is from 16,000 to 27,000 RPM.

PARTS

The four major parts of the router are the motor unit, base, hand knobs, and depth adjustment ring. The motor is self-contained and determines the RPM of the cutter. The base is the platform that rides the surface of the wood and controls the depth of cut. The two hand knobs or handles are conveniently located for easy grasp and control; some routers have a handle and a knob, in which case the trigger switch is located on the handle. The depth adjustment ring sets the depth of the cut and is usually fastened to the base.

BITS AND ACCESSORIES

Router bits come in sizes and shapes that permit almost any cut desired. (See Figure 2–4(a)–(b).) Router bits fall into two major classifications: (1) one-piece bits, which have a shank built into the cutting head and which fit into the collet (chuck) of the router motor; and (2) screw-type bits which have an arbor and a pilot. The pilot is screwed into the bottom of the arbor and controls the horizontal depth of the cut by riding along the edge of the work-plate.

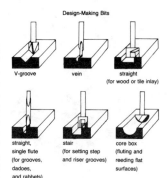

Design-Making Bits

V-groove

vein

straight
(for wood or tile inlay)

straight, single flute
(for grooves, dadoes, and rabbets)

stair
(for setting step and riser grooves)

core box
(fluting and reeding flat surfaces)

(Figure 2-4b)   Router bits (con't)

An important router accessory is an edge guide, which can be used in cutting an edge, radius, groove, or circle. (See Figure 2–5.)

(Figure 2-5)   Router guide

Some router manufacturers make shaper stands in which the router can be inverted and secured. Router stands allow the router to be set for a specific depth and type of cut. This is very handy if you are making a number of repeated runs that must be uniform, as for the molding of a custom job.

Another important mechanical aide is the hinge-built template, which is used as a guide in cutting the notches in doors and frames for butt hinges.

Dovetail templates are available for specialized cabinet work. These allow the operator to layout and cut dovetail joints for drawers.

Basic Procedures for Using Routers

1. *The router is most commonly used for cutting decorative edges on wood. When making cuts on all four edges of a board, make the*

first cut on the end across the grain. If any chipping occurs at the end of a cut, remove it by making the next cut parallel to the grain.

2. Select the proper bit.

(Figure 2-6)   Adjusting a router

3. Fasten the bit in the chuck, following the manufacturer's directions. If your router has a locking device to prevent the motor shaft from turning, use a wrench to tighten or loosen the collet nut. If there is no locking device, you must use two wrenches: one to hold the lower nut and the other to tighten or loosen the upper nut. (See Figure 2–6.)

4. Adjust the base for the correct depth of cut. Follow the manufacturer's instructions, making sure that the router is first placed on a flat surface. Generally, the wing nut should be loosened and the collar should be turned until the bit just touches the surface. Lift the router and turn the collar counterclockwise until the bit is lowered to the desired depth, at which point the nut should be tightened.

5. Make a test cut on a piece of scrap wood.

6. Fasten the workpiece firmly.

7. Place the router base on the board with the bit over the edge.

8. Turn on the switch. Push or pull the router against the edge of the board until it hits the bit or the cutter collar.

9. Push or pull the router from left to right.

10. Finish shaping the edge.

11. Turn off the power. Remove the router from the board or project.

12. Consult the manufacturer's manual for specific instructions on how to make such specialty cuts as dadoes, veins, and grooves. The method of operating the router remains the same in every job—move the cutter slowly into the board or project while remaining constantly aware of the cutting edge's position.

FACTORS TO CONSIDER

- Tungsten-carbide-tipped bits retain sharpness under normal use, but high-speed bits dull quickly when cutting hardwoods or woods with high moisture content.

- Listen to the sound of the motor when engaging the cutter into the wood. The proper setting for cutting will cause only a slight reduction in the motor speed.

- Don't move the cutter too slowly, or you will burn the wood and ruin the bit.

- Moving the router too fast will slow the motor down too much and cause overheating.

- Practice running the router until you have developed a sense of what proper cutting sounds like. You will soon learn that one sound may mean overheating, and another burning wood.

- Never rout wood that has hardware, such as nails or screws, in it. Hitting hardware with the router is dangerous and can ruin the cutter.

- It is better to take several light passes (cuts at a depth shallower than the desired finished cut) than to try to remove all excess stock at once. By taking off too much stock you may overheat the cutter and motor and split the stock.

SAFETY PROCEDURES FOR USING ROUTERS

Eye protection. Wear goggles or a face shield when using the router.

Ear protection. Wear ear plugs to save hearing from damage caused by the sound of the router's high-frequency motor.

Grounding. Check to see that the electrical connection is grounded.

Electrical power. Disconnect the motor unit from the electrical power outlet when you change bits, cutters, or attachments.

Stock. Be certain the work is securely clamped so it will remain stationary during the routing operation.

Starting. Place the router base on the work, template, or guide, with the bit clear of the wood, before turning on the power. Hold the router firmly when turning on the motor, since starting torque could wrench the tool from your grasp.

Direction of movement. Move the router from left to right when you cut straight edges. Move it from right to left when you cut circular or curved edges.

Stopping. When the cut is completed, turn off the motor. Do not lift the machine from the work until the motor has stopped.

Adjustments. Always unplug the motor when mounting bits or making adjustments.

SANDERS

The portable power sander saves time and effort in smoothing surfaces. Carpenters usually include three basic types of sanders in their power tool collection. These are belt, disc, and finish sanders.

BELT SANDER

The belt sander has an abrasive belt that runs continuously over pulleys that are situated at both ends. The average portable model weighs between 10 and 20 pounds. Its size is usually determined by the size of the belt. Two common sizes are the 3-by-24-inch belt and 4-by-27-inch belt. The heavier duty belt sanders are equipped with dust bags.

PARTS

The main parts of the belt sander are listed below along with explanations of their functions.

Motor housing. The motor housing, which encloses the motor, is usually made of cast aluminum to reduce the weight of the tool. It is located on the front portion of the frame.

Frame. The frame is an aluminum casting to which is fastened the motor housing and other parts of the machine.

Traction wheel. The traction wheel is the rear pulley that drives the abrasive belt.

Idler wheel. The idler wheel is the front pulley. It provides tension on the abrasive belt.

Handle. The handle is usually made of composition or plastic (shockproof) material; it contains the trigger switch. A ball, knob, or front handle is also mounted because operation of the sander requires the use of both hands.

Tracking adjustment knob. The tracking adjustment knob is used to align the belt on the wheel. After the belt is put on the sander, the sander is tilted back and turned on. The tracking adjustment knob is turned to move the belt in or out until it runs centered on the wheels. (See Figure 2–7.) The new top-of-the-line belt sanders are self-tracking and require no adjust-ment—saving both time and wear on the belt and the sander.

(Figure 2-7)   Belt sander

Dust bag. This is a removeable bag that catches the dust and that can be emptied when full.

Shoe base. The shoe base is a metal plate over which the back of the abrasive belt moves. It forces the belt to make contact with the wood.

Abrasive belt. Belts are available in many grits (rough to fine) to fit different types of portable belt sanders. See the section on abrasives for a discussion on abrasive selection.

Basic Procedures for Using Belt Sanders

1. *Carefully read the manufacturer's directions for installing the abrasive belt. Follow them for best results.*

2. *The key to successful operation of the belt sander is proper belt tracking. To set the belt for proper tracking, tilt the sander back, start the motor, and turn the belt alignment screw so the outer edge of the belt runs even with the ends of the wheel.*

3. *Hold the sander over the work. Start the motor, and then lower the sander carefully and evenly onto the surface.*

4. *For fast wood removal, operate the sander at about a 45-degree angle to the grain direction, using a fairly coarse belt.*

5. *When using any sander, always go with the grain—never across it. Failure to sand with the grain will result in permanent marks in the wood.*

6. *Move the sander back and forth over the surface in even strokes.*

7. *At the end of each stroke, shift it sideways about half the width of the belt.*

8. *Keep the sander moving when the motor is on so a groove will not be cut in the wood.*

9. *Hold the sander level while in operation and do not press down on it. The weight of the sander is sufficient to provide proper pressure for the sanding action.*

10. *When the work is completed raise the sander from the surface and allow the motor to stop.*

DISC SANDER

The disc sander comes in two forms: the rubber disc and the disc sander.

The rubber disc is actually an accessory for the portable electric hand drill. It is attached by means of a metal rod that extends from one side. The rod is secured in the drill chuck in the same manner as a bit. The abrasive disc is fastened to the rubber disc by means of adhesive or by a large washer and nut that is secured to the center of the disc. The rubber disc sander is used primarily for sanding uneven surfaces or hard-to-reach areas where the rubber disc can conform to the shape of the object being sanded.

The second type of disc sander is a self-contained power unit that is used primarily for sanding stair treads, along baseboards, in corners, inside closets, and in other places that are inaccessible to larger sanders.

PARTS

The main parts of the disc sander are described below.

Motor housing. The housing is usually made of aluminum alloy, which is durable and lightweight. It houses the motor and all other parts of the sander.

Handles. The handles are made of plastic and are located on either side of the motor housing. They are used to control and guide the sander. The trigger switch is located in one of the handles.

Spindle. This is the threaded mechanism that is driven by the sander motor and into which the lock nut is screwed to hold the abrasive disc.

Dust bag. This is a removable bag that catches the dust and that can be emptied when full.

Abrasive disc. Abrasive discs come in different sizes and grits. Both size and grit selection depend on the type of disc sander and the purpose for which it is used. See the section on abrasives for a discussion of abrasive selection.

Basic Procedures for Using Disc Sanders

1. *Carefully read the manufacturer's directions for installing the abrasive disc. Follow these directions for best results.*

2. *Hold the sander over the work. Start the motor, and then lower the sander evenly onto the surface.*

3. *Go with the grain if at all possible. In smaller, out-of-the-way locations where this is not possible, put on a finer disc and proceed lightly to prevent cross-grain scoring marks.*

4. *Move the sander back and forth over the surface in even strokes.*

5. *At the end of each stroke, shift the sander sideways about half the width of the disc.*

6. *Keep the sander moving when the motor is on to prevent cutting a circle in the wood.*

7. *Hold the sander level while in operation and do not press down on it. The weight of the sander is sufficient to provide proper pressure for the sanding action.*

8. *When the work is completed, raise the sander from the surface and allow the motor to stop.*

FINISHING SANDER

Finishing sanders have either an orbital (circular) or oscillating (vibrating) motion. The two types look similar.

The designation "finishing sander" means that this tool is used for fine sanding. It is sometimes used to obtain a finer surface finish after using the belt sander. The orbital type cuts faster because of the way it moves its abrasive pad.

Finishing sanders vary in size and take one-fourth, one-third, or one-half sheets of abrasive paper or cloth. Their weights vary from 4 to 10 pounds.

PARTS

The basic parts of a finishing sander are described below.

Motor housing. This housing is usually made of cast aluminum or plastic to reduce the weight of the tool. It encloses the motor and all other parts of the sander.

Pad plate. Abrasive sheets are fastened and locked to the pad plate, which forces the sheets to make contact with the wood.

Clamps. The pad plate has a clamp at each end that holds the abrasive sheet.

Handle. The hand-fitting aluminum or plastic handle includes the trigger switch. There is usually a second handle on the front so that this tool can be operated with two hands.

Abrasive. Finishing sanders use cut-to-fit portions of standard abrasive sheets. The size of the cut-to-fit abrasive strip depends on the size of the sander. See the section on abrasives for a discussion of selections.

Basic Procedures for Using Finishing Sanders

1. *Select an abrasive sheet with a grit that is suitable for the type of work to be performed.*
2. *Carefully read the manufacturer's directions for installing the sandpaper. Practically every manufacturer has its own method of attaching the sandpaper to the operating end. Some use clamps, and others use a knurled rod or a sliding clip arrangement.*
3. *Connect the plug to the electrical power outlet.*
4. *Lift the sander off the board. Start the motor.*
5. *Set the sander down evenly on the board. Move it back and forth.*
6. *Guide the sander with the handle. Use both hands until you get the feel of its operation. The weight of the machine itself exerts sufficient pressure for most sanding.*
7. *The finishing sander is effective when more than one grain direction*

is encountered, such as with miter joints, because the sander has a small circular motion that will not scratch the wood when moving in the opposite direction of the grain.

8. *Turn off the power. Change the abrasive sheets. Continue with finer grits until the surface is smooth.*

FACTORS TO CONSIDER

■ Portable sanders, with proper abrasive materials, can be used on metals, slate, marble, or plastics. The only difference in working on these materials is that there is no grain to consider.

■ Use a spaced, coarse-grained, or open abrasive belt or sheet to remove old paint, enamel, varnish, and lacquer from flat surfaces. When removing finishes, lower the sander at the far end of the work and pull it back. Raise the sander, and repeat the cycle in another area.

■ When starting to operate a belt sander, make sure your footing is secure. When the sander is first lowered onto the material, it has a tendency to pull you forward.

■ Check a belt sander often for proper tracking of the belt. If the tracking is off and the belt goes toward the housing of the sander, it will cut grooves in the housing and can ruin it. If the tracking drives the belt away from the housing, the belt will fly off and the traction wheel will scar the work surface.

■ Check any sander you use to make sure the locking trigger is off before it is plugged in. If the trigger is locked while you are plugging in the power cord, the sander can run off the work area, breaking the housing, wheels, etc.

■ Wear a dust mask and goggles when operating a sander.

■ Sanders are safe and easy to use, but if your fingers make contact with a moving belt, you will find out how abrasive even a smooth grit can be.

DRUM SANDER

The drum sander is not one that most carpenters own since it has a specialty use and can be rented for a reasonable rate at most tool rental stores. It is used primarily in refinishing hardwood floors.

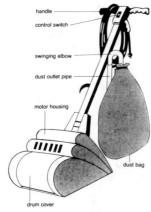

(Figure 2-8)   Drum sander

The drum sander consists of a revolving rubber-covered drum mounted on a wheeled frame that tilts backward to lift the drum from the floor. A sheet of sandpaper is wrapped around the drum; a removable bag attached to a vacuum arrangement collects the dust. (See Figure 2–8.) A drum sander with a 1½ HP motor that operates on standard current is appropriate for most work.

Before sanding floors, always make sure that any nails have been set far enough below the surface so as to not be exposed during the sanding operation. Nail heads can tear the sandpaper and ruin the entire sheet.

The drum sander tends to pull away from the operator, so be prepared to hold it back so that it moves at a slow and even pace.

Basic Procedures for Using Drum Sanders

1. *Select the appropriate grit sandpaper. Sandpaper for the drum sander can usually be purchased at the rental site.*

2. *Follow the manufacturer's directions on installing the sandpaper.*

3. *Tilt the machine backward until the drum clears the floor, and then turn on the power and gently lower the drum till it touches the surface.*

4. *The drum sander operates in both forward and backward directions. Raise the drum at the beginning and end of each pass across the floor.*

5. *Sand with the grain whenever possible.*

SAFETY PROCEDURES FOR USING SANDERS

Hands. Keep both hands on the handles of the belt sander. This makes it impossible to touch the sanded surface while sanding and endanger your hand. This rule also applies to the drum sander. By keeping both hands on the handles, you will be able to control the sander so it will not jerk away and cause damage to the work area or someone in it.

Ear and face protection. Wear ear plugs and a dust mask.

Electrical grounding. Check to see that the electrical connection is grounded.

Electric cord. Arrange the electric cord so that it cannot be caught by the abrasive belt. A good arrangement for safety is to hang the cord over your shoulder. Keep the cord free, and prevent it from being drawn between the abrasive belt and the housing.

Abrasive belts and sheets. Make certain that you select the correct size belt or sheet and the proper grade of grit to do the job.

Changing belts and sheets. Disconnect the plug from the power outlet when changing abrasive belts or sheets.

Starting. Always lift the sander before starting to sand and also before cutting off the power. When starting or stopping the drum sander, always tilt it backwards until the drum is clear of the work surface.

Sander at rest. Lay the belt sander on its side when you are not using it. When not using the drum sander, unplug it. If there is concern about damage to a finished surface, set the drum sander on a piece of cardboard to protect the surface.

Stopping. Shut off the power and do not put down the portable sander until it comes to a complete stop. The drum sander should be tilted backwards until it comes to a complete stop. Then it can be slowly set on the work surface.

SAWS

Saws are made to cut materials. The simplicity of this statement is deceiving, as saws come in many varieties and are designed for different uses. The saws discussed in this section are those most commonly used in basic construction.

CIRCULAR SAW

The portable circular saw is one of the most extensively used power tools. A quality circular saw has built-in balance that allows an operator to use it for hours without excessive fatigue.

The size of the circular saw is determined by the diameter of the largest blade it will take. Most carpenters prefer a 7- or 8-inch saw. Fitted with proper blades or with abrasive discs this saw can be used to cut many materials. It slices ceramics, slate, marble, tile, nonferrous metals, corrugated galvanized sheet steel, and almost any other kind of building material. It is also useful for cutting grooves, dadoes, and rabbets. Circular saws weigh from 6 to 12 pounds, with the horsepower rating on the motor ranging from $\frac{1}{16}$ to $1\frac{1}{2}$.

PARTS

The basic parts of the circular saw are described below.

Base plate. The base plate supports the motor, handle, blade, and the entire mechanism of the tool. It is adjustable, controlling the depth and angle of the cut.

Handle. The handle is contoured to give balance in operation. Most handles have a self-contained, insulated, safety trigger switch.

Motor. The motor is contained in a body housing that also includes the gear drive that turns the blade. This housing protects the motor from dust and possible damage from handling.

Retractable safety guard. This spring guard lowers to cover the saw teeth when the tool is not in operation. It retracts exposing the teeth, when the tool is in use.

Bevel adjustment. A bevel scale is usually found on the front of the tool. It is used to set the saw blade at any angle from 45 to 90 degrees to the base plate. The adjustment is secured with a tilt lock knob.

Blade. The rough-cut combination blade is popular because it is suitable for both ripping and crosscutting. Some carpenters prefer carbide-tipped teeth because they usually stay sharp longer than those of a standard blade. A hollow ground blade is used when the cut must be very smooth. For prolonged crosscutting or ripping, a regular crosscutting or ripping blade should be used.

Basic Procedures for Using Circular Saws

1. *Select the proper blade for the cutting operation.*
2. *Place the blade on the motor shaft, or arbor, as specified in the manufacturer's instructions, and secure it.*

3. *Adjust the depth of the cut so that the blade saws just through the board or so that one full tooth penetrates the work. (See Figure 2–9.)*

4. *Plug the cord into an outlet.*

5. *Rest the base on the work and align the guide mark with the layout line.*

(Figure 2-9)   Adjusting a circular saw

6. *Back the saw slightly away from the work. Turn on the switch and allow the motor to reach full speed.*

7. *Move the blade steadily forward through the board. Do not force it. Move the saw only fast enough to keep it cutting.*

8. *Release the switch when the cutting is completed. Hold the saw until the brake has stopped the blade.*

9. *When ripping, you can free-hand the cut or, for greater accuracy, attach and use the rip guide. The inside of the rip guide should be waxed so that it runs smoothly along the edge of the work. (See Figure 2–10.)*

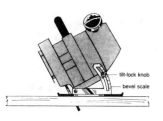

10. *When making rip cuts wider than the rip guide, a board can be clamped to the stock and used as a guide by placing the saw's base plate against it.*

(Figure 2-10)   Using a circular saw

Pocket Cut

Pocket cuts (cuts made inside the area of the material) are often made on the job. To complete these cuts accurately follow the steps below.

1. *Plug the cord into the outlet.*

2. *Start near the corner of one side by placing the front edge of the base plate firmly on the work.*

3. *Hold the saw up so that the blade clears the material. Be sure you have adjusted the blade for the proper depth of cut.*

4. *Push the retractable lower guard all the way back so that the blade is exposed.*

5. *Start the motor, and lower the blade into the board. The front of the tilting base serves as a pivot as the blade is lowered.*

6. *Follow the marked line right up to the corner.*

7. *Use the procedure described above for sawing the other lines of the pocket cut.*

8. *Use a compass saw to cut the corners cleanly and accurately.*

Bevel Cut
1. *Set the blade at the desired angle to the base, and tighten the tilt-lock knob.*

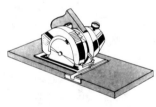

2. *Adjust the depth of cut so that the blade cuts just through the piece.*

3. *Follow standard circular saw operations to complete the cut. (See Figure 2–11.)*

Grooves, Dadoes, and Rabbets

These cuts involve the same basic operation of the saw, except that one depth is set according to the needed cut and the clamped guides direct the saw in making it.

(Figure 2-11)   Cutting a bevel

SAFETY PROCEDURES FOR USING CIRCULAR SAWS

Grounding. Be sure that the electrical connection is grounded.

Electric power. Check that the trigger switch is in the "off" position before connecting the saw to the power source.

Eye protection. When operating the saw, wear safety glasses.

Secure workpiece. Be sure that the workpiece is held securely to the work surface.

Blade. Always use a sharp blade that is properly set. Study the operator's manual for instructions on changing the saw blade.

Cutting depth. Always check that the depth adjustment is correct before cutting. Be careful not to cut into the sawhorse or other supporting devices.

Adjustments. Check that the base and angle adjustments are tight. Always unplug the saw to change blades or make adjustments.

Starting. Always place the saw base on the stock with the blade clear before turning on the switch.

Cutting. While cutting, stand to one side of the cutting line. A saw meeting unexpected resistance (such as a knot) can fly back with considerable force. Never reach under the material being cut.

Turning off power. Always turn off the power switch once the cut is completed. Do not take your hands off this tool until the motor and blade have stopped.

FACTORS TO CONSIDER

- Make sure all guards are maintained and in place.

- As with using other power tools, make sure the materials being worked on are free from hardware (nails and screws) that might damage the blade or cause injury.

- The bottom or base plate of the saw should be kept clear of gum, dirt, and pitch. This can be done by rubbing down the base plate with turpentine. Waxing the base plate will allow it to move freely.

MITER SAW (CUTOFF SAW)

The miter saw is a specialty tool that is becoming an essential part of a carpenter's power tool assortment because of the speed and accuracy of its cuts. The miter saw is a particularly useful tool for home remodeling projects, such as building decks and cutting framing and molding for room additions.

This saw is basically a portable circular saw attached to a miter box. The saw is enclosed in a spring-loaded holding device that allows the operator to set the saw at any desired angle of cut, ranging from 45 to 90 degrees. The size of most miter saws is 7 inches, and the most commonly used blade is a hollow ground combination blade. The miter saw is often called the cutoff saw because on many job sites it is set up and used primarily to cut boards to length. The miter saw also has a capacity for tilting to make compound miters.

Basic Procedures for Using Miter Saws

1. *Set the saw to the desired angle of cut. A gauge located at the bottom front of the miter box is used for setting angles.*

2. *Place marked stock against the back fence of the miter box. Hold the stock firmly.*

3. *The saw switch is located in the miter box handle for easy access. Squeeze the trigger switch to turn the saw on, and let the motor reach full speed.*

4. *Grasp the miter box handle and lower the saw until the blade has cut through the stock.*

5. *Release the trigger switch while allowing the miter box handle to return to an upright position. Do not let go of the handle; let spring pressure lift the saw.*

6. *The brake on the saw will stop the blade almost instantly. When the blade has stopped moving, remove the stock.*

SAFETY PROCEDURES FOR USING MITER SAWS

Guards. Keep guards in place while operating.

Grounding. See that the electrical connection is properly grounded.

Eye protection. Wear safety glasses or a face shield to protect eyes from sawdust and other debris.

Hands. Keep hands clear of the cutting area at all times. Never reach in to get the stock before the blade has stopped rotating.

Angles. Lock the saw securely at the angle of the cut.

Stock. Hold the stock firmly against the fence.

Brake. Make sure the brake is well maintained and in good working order at all times.

Blade. Always use sharp blades with the correct set.

RADIAL ARM SAW

The radial arm saw is a precision machine that is capable of doing a wide variety of operations. It gets its name from the arm that can be rotated 360 degrees to the right or left. It is a versatile tool that is used for numerous operations on the construction site. Radial arm saws are normally set up in a central location on the work site, and work is brought to the saw. Often the radial arm saw is mounted on a trailer so that it can be easily transported from one site to another. In the home workshop, the radial arm saw may be the only stationary power saw needed. It rips and crosscuts, cuts dadoes, miters, compound miters, and rabbets for framing, furniture making, and many other jobs.

Radial saw sizes are measured by the blade diameter and vary from 8 to 20 inches. The 10- to 14-inch saws are the most common on the construction site. The blade is attached to the motor by a direct-drive motor arbor, or shaft. The motor is mounted in a yoke and may be tilted for angle cuts. The yoke is suspended from the arm on a pivot that permits the motor to be rotated in a horizontal plane. Adjustments make it possible to perform many sawing operations. Motor speed on most models is between 3,425 and 3,450 RPM. The surface speed at which the material can be fed into the saw is figured in surface feet per minute (SFM) and varies with the diameter of the blade. This speed can be computed by multiplying the motor RPM by the blade circumference in inches, and then dividing by 12.

PARTS

The parts of a radial arm saw are described below.

Table. The table is a series of boards held in place with guide fence controls.

Guide fence. The guide fence provides backing for crosscutting, ripping, and many other operations. This wooden fence guide is removable. It can be positioned between any of the table boards, depending on the width of the material to be cut.

Base. The base is usually made of steel. It supports the table and the radial arm.

Yoke. The yoke holds the electric motor, which hangs from the track arm. The yoke can be moved along the horizontal track arm or held stationary at any point. The blade and other tools are fastened to the motor shaft, or arbor.

Track arm. The track arm enables the yoke to move back and forth. It is adjustable and can move 360 degrees to the right or left.

Overarm. The overarm is the top horizontal bar that controls the track arm. It is supported by the column at the rear. This arm raises or lowers the cutting blades.

Upright column. The upright column is the steel cylinder that provides support for the overarm at the rear. It holds the radial arm mechanism above the table and base.

Guards. The blade guard protects the operator from the blade and cutters. The antikickback attachment has fingers in front of the blade guard that keep boards from kicking back during ripping operations.

Controls and scales. The saw has several controls based on standard units of measurement that are used for determining the width and angle of cuts. The scales are used to measure depth of cuts.

SAW BLADES AND ACCESSORIES

The blades used in a radial arm saw are of the same type as those used in portable circular saws: rip, crosscut, and combination blades. Due to the time it takes to change blades, the combination blade is the most commonly used. Since the radial arm saw is most often used when cutting compound miter joints, the hollow-ground combination blade is the one most often chosen for smooth, even cuts.

Attachments for the radial arm saw include the dado head, molding head, shaper cutters, router bits, rafter-head cutter, and sanding discs and drums. All of these attachments can complete a wide range of operations on the job site. For a complete listing of the functions of these accessories and how they can best be used, consult the manufacturer's manual for each of these devices.

Basic Procedures for Using Radial Arm Saws

1. *Read the manufacturer's manual carefully for specific operating procedures for your saw.*
2. *Observe all safety rules for the operation of the saw.*
3. *Select the proper blade for the cutting operation.*

Crosscutting

1. *Set the track arm at zero on the miter scale. Lock the setting securely.*

2. *Set the depth of cut by turning the elevating crank. The teeth of the saw blade should barely scratch the top of the table if you wish to cut entirely through the stock.*

3. *Place the stock on the tabletop against the guide fence.*

4. *Turn on the power switch. Make sure the saw blade is behind the guide fence.*

5. *Pull the yoke handle slowly as the saw cuts across the board. Hold the stock firmly against the guide fence. To cut more than one piece to the same length, clamp a stop to the guide fence.*

Ripping

1. *Set the track arm to the crosscutting position.*

2. *Adjust the yoke so that the blade is parallel to the guide fence.*

3. *Set the depth of cut by turning the elevating crank.*

4. *Push the carriage along the track arm to the ripping width desired. Lock it in place with the rip-clamp handle.*

5. *Place the board on the tabletop against the guide fence.*

6. *Lower the infeed end of the guard to clear the stock; lock it.*

7. *Adjust the antikickback guard. Make certain that the points of the fingers are set about ⅛ inch below the surface of the board.*

8. *Turn on the power switch.*

9. *Feed the board into the saw slowly. Make sure that it moves against and along the guide fence. Do not feed the stock into the antikickback end of the saw guard. (See Figure 2–12.)*

10. *Whenever making a rip cut on a radial saw, use a push stick to push away the stock that has been cut.*

(Figure 2-12) Sawing stock

Cutting a Miter

1. *Loosen and swing the track arm to the desired right- or left-hand angle on the miter scale. Lock it securely.*

2. *Set the depth of the saw cut by adjusting the elevating crank. The cutting edge of the blade should barely touch the table. This will cut completely through the stock.*

3. *Place the stock on the tabletop against the guide fence.*

4. *Turn on the power switch.*

5. *Pull the saw slowly across the board, as in crosscutting. It is always a good idea to check the setup by cutting a piece of scrap stock first.*

6. *Return the saw to its position behind the guide fence. Turn off the switch. Always do this before removing the stock.*

Crosscutting a Bevel

1. *Raise the motor and the cutting head with the elevating crank just high enough to allow the blade to be tilted.*

2. *Adjust the track arm and yoke for crosscutting; then tilt the blade to the desired angle on the bevel scale.*

3. *Set the depth of cut by turning the elevating crank.*

4. *Place the board on the tabletop and against the guide fence.*

5. *Turn on the power switch.*

6. *Pull the saw slowly across the board as in crosscutting.*

Ripping a Bevel

1. *Adjust the machine for straight ripping.*

2. *Position the carriage on the track arm for proper tilt of the bevel.*

3. *Rip the angle in the stock by following steps 5 through 10 under "Ripping."*

SAFETY PROCEDURES FOR USING RADIAL ARM SAWS

Instruction manual. Read the instruction manual carefully before you attempt to use the radial arm saw. Each manufacturer has slightly different instructions on how to make adjustments and perform the various processes on the machine.

FACTORS TO CONSIDER

- Keep all guards well maintained and in place.

- As with other power tools, make sure the materials being worked on are free from hardware (nails and screws) that might damage the blade or cause injury.

- The radial arm saw has a real use on the job site when cutting large pieces of lumber that are difficult to slide across a saw table.

- The manufacturer's manual illustrates the numerous operations that can be performed with the radial arm saw. It's important to study the manual carefully so that you will know the machine's total operating capacity.

- The radial arm saw has many moving parts that must be adjusted for different operations. If these adjustments are not made or if makeshift tables and fences are used in conjunction with this tool, it will be impossible to make accurate cuts.

Eye protection. Wear safety goggles to protect eyes from flying sawdust and wood chips.

Sharp blades. The teeth of the saw blades and the cutter knives should be kept sharp at all times. Be sure that the teeth of the saw blades or cutters point in the direction of the arrow on the saw guard.

Holding stock. Stock must be held firmly on the table and against the fence for all crosscutting operations. The ends of long boards must be supported level with the table.

Guards. Keep the guards and antikickback device in position.

Adjustments. All operating adjustments should be made and locked before starting the machine.

Push stick. Use a push stick when ripping and cutting grooves and rabbets on narrow stock even though your fingers may fit between the blade and the fence.

Saw pull. Remember that this saw pulls itself into the work. It is necessary to push back on the handle to prevent the saw blade from cutting too fast and choking.

Saw position. Always return the saw to the rear of the table after completing a crosscut or miter cut. Never remove stock from the table before the saw has been returned.

Distance from blade. Throughout the cutting operation, keep your hands 6 inches away from the path of the saw blade.

Removing stock. Shut off the motor and wait for the blade to stop before removing the workpiece.

Leaving the machine. Do not leave the machine until the blade has stopped rotating.

Work area. Keep the table clean and free of scrap pieces and excessive amounts of sawdust. Do not attempt to clean off the table while the saw is running. Make sure sawdust is swept off the workpiece, because the hand holding it against the fence can slip on the sawdust toward the moving blade.

SABER SAW (JIG SAW)

The saber saw is also called a portable jig saw or bayonet saw. It is used for a wide range of light work by carpenters, especially when making irregular or close-work cuts such as those for electrical boxes. The saber saw weights between 3½ and 4½ pounds. The cutting speed is approximately 4,200 strokes per minute (SPM).

The different saber saw models have slight variations with respect to the way in which blades are mounted in the chuck. Follow directions in the manufacturer's manual. Many specialty blades are available for the saber saw, and these can be used to cut a variety of materials. Blades for woodcutting have from 6 to 12 teeth per inch. For general-purpose work, a blade with 10 teeth per inch is satisfactory. Always select a blade that will have at least two teeth in contact with the edge being cut.

PARTS

The basic parts of the saber saw are the body, handle, and base. The body is a lightweight metal alloy housing that contains the working mechanism. The handle is the sturdy piece on top that allows the operator to manipulate the saw with ease. It usually includes a trigger switch for turning the saw on and off. Some models have an auxiliary handle on the side. A few of the lighter saber saws are built so that the handle is actually a part of the body. The base of the saw serves as an inverted table. It provides a surface to guide the saw on the work. On most models it can be tilted.

Basic Procedures for Using Saber Saws

Ripping and Crosscutting with a Guide

1. *Fasten a blade on the saw that is suitable for the material to be cut.*
2. *For ripping or crosscutting narrow stock, fasten the guide in place and adjust it to the desired width. If the piece to be cut off is wider than the guide permits, clamp a wooden fence on the board.*
3. *Hold the workpiece so that the blade can cut through freely.*
4. *Start the motor. Slowly saw the workpiece. Keep the guide firmly against the edge of the stock's end, or keep the saw against the wooden fence or guide.*

Sawing Circles

1. *Position the workpiece so that it is securely against sawhorses or some other device that allows space underneath for the blade to pass through.*
2. *Drill a ³/₁₆- or ¹/₄-inch starting hole through the workpiece. This should touch the circle layout line on the waste side of the line.*
3. *Attach the correct blade to the saw.*
4. *Drive the circle guide pin into the workpiece at the center of the circle.*
5. *Turn the rip guide over and fasten it in the saw to the radius desired.*
6. *Start the motor with the saw in place. Slowly push the saw as it makes the circular cut.*

SAFETY PROCEDURES FOR USING SABER SAWS

Planning. Lay out and carefully plan your work before sawing.

Eye protection. Wear safety goggles while operating the saw.

Blades. Make sure you have selected a blade that is the correct type and size for the job to be done. Also make sure it is correctly mounted.

Adjustments. All adjustments must be secure before starting this tool. Disconnect the saw to change blades or make adjustments.

Electrical grounding. Be sure that the electrical connection is grounded.

Holding material. Hold or clamp the stock to be cut so that it cannot vibrate. Be sure that the work is well supported. Check the clearance below the workpiece so that you do not cut into sawhorses or other supports.

Contact of saw. Place the base of the saw firmly on the stock before starting the cut. Turn on the motor before the blade contacts the work.

Cutting curves. Do not attempt to cut curves so sharp that the blade will be twisted.

Stopping. Once you have turned off the power switch, be sure to hold the saber saw until all motion stops; then you may safely return the tool to its proper place.

FACTORS TO CONSIDER

- One of the uses of the saber saw is to cut out sink openings in laminated countertops. To perform this operation, a plunge cut is used. To start the cut, place the forward edge of the saw base firmly on the edge of the material so that the saw is tilted. Do this only when the thickness of wood or plastic laminate doesn't exceed 1 inch. The plunge cut saves the time of drilling a starter hole for the saw blade.

- The saber saw is very easy to operate. After a little practice, good results can be achieved.

- Keep in mind the length of the saw blade that goes through the workpiece. Be sure that the workpiece is positioned to allow clearance under the stock so that the blade will not hit or mar anything below the workpiece.

- In the course of operating a saber saw, blades will be broken. To avoid the frustration of not having enough blades to finish a job, always keep a good supply of them in your tool kit.

TABLE SAW

The table saw is often referred to as a bench saw or circular saw. It can, with the proper accessories, perform a variety of cutting operations including: constructing built-in cabinets and units, and cutting and fitting moldings and other inside trim. Some models of this machine are portable. The motor and table assembly can be easily separated from the base so that the machine can be transported to a central location at the work site. These models are known as contractor saws.

The size of a table saw is measured by the diameter of the largest saw blade it will accommodate. Generally, the 8- to 10-inch saws are used. The cutting speed is approximately 8,100 SFM at approximately 3,100 RPM.

The blades used on the table saw are basically the same as those used on the radial arm saw. There are rip, crosscut, and combination blades for general use. The combination blade is the most convenient blade to use. The operator does not have to take time to change the blade each time a change is made from ripping to crosscutting and vice versa. For the smoothest cut, the hollow ground combination blade should be used. A carbide combination blade is durable and is most often chosen when doing a lot of general construction cutting.

PARTS

The major parts of a table saw are described below.

Arbor. The arbor has a shaft that holds the blade, dado head, or molding head.

Frame. The frame includes the base, the housing for the internal operating mechanisms, and the motor.

Table. The table supports the rip fence, cutoff guide, safety guard, and the wood stock being cut.

Rip fence. The rip fence is used as a guide for ripping. It often has a vernier adjustment for fine calibrations.

Miter gauge cutoff guide. The miter gauge cutoff guide is used as guide for crosscutting.

Saw-tilt hand wheel. This is a hand wheel for tilting the saw blade or the table, depending on the type of machine.

Blade-raising wheel. This is a hand wheel that regulates the cutting height of the saw blade.

Safety guard. The safety guard protects the operator from the saw. It is detachable, but it should be used whenever possible. The safety guard has antikickback pawls with teeth that grab the workpiece as it passes by and prevent it from being thrown back toward the operator.

Splitter. The splitter is used to prevent the stock from binding against the saw blade while the cut is being made. Sometimes, this is a separate piece, but it is often part of the guard.

Electric Motor. The electric motor drives the circular saw.

Basic Procedures for Using Table Saws

Crosscutting

1. *Place the cutoff guide in the slot or groove on the table. Usually, the left groove is more convenient for crosscutting.*

2. *Check that the cutoff guide is set at the correct angle of 90 degrees for a right angle by using a try square held against the blade and miter gauge.*

3. *Be sure that the saw blade is set to cut at 90 degrees from the table edges.*

4. *Hold the board firmly against the cutoff guide. Place the board so that you saw on the waste side of the marked line.*

5. *Start the saw, and let it reach full speed.*

6. *Push the cutoff guide and the board forward, holding the board firmly against the guide. The saw blade should extend above the board approximately ¼ inch.*

7. *Pull back the board and the cutoff guide at the end of the cut.*

8. *When several short pieces are to be cut to the same length, clamp a block on the rip fence and position it to act like a length gauge. This gives sufficient clearance between the saw blade and the rip fence to keep the piece from sticking between them. This is the only time the rip fence and miter gauge should be used together; otherwise, the workpiece will get trapped between the blade and fence and will be kicked back.*

Ripping

1. *Adjust the height of the rip or combination blade to approximately ⅛ to ¼ inch above the thickness of the stock.*

2. *Fasten the rip fence at the desired distance from the saw blade. Also, make sure the fence is at a right angle to the tabletop.*

3. *Remeasure the width of the rip cut to see that the rip fence has not slipped.*

4. *On specific cuts of close tolerance, such as stock used for cabinet*

fronts, make a trial cut on a piece of scrap stock to check the accuracy. Allow at least 1/16 inch for planing with a jointer or a hand plane.

5. Turn on the switch; allow the motor to reach full speed.

6. Place stock firmly on the tabletop. Press it against the rip fence.

7. Push the stock with steady pressure. Keep the safety guard in place, and whenever possible use the antikickback fingers and the splitter attachment. The splitter holds the saw cut open, eliminating binding.

8. To rip pieces less than 4 inches wide, use a push stick for safety. Sometimes when thick stock is being ripped, it is wise to set the saw to cut only part of the thickness the first time. Reset it for a deeper cut; and if necessary, make several cuts to complete the ripping. A splitter cannot be used until the last cut when following this procedure. (See Figure 2-13.)

(Figure 2-13) Making rip cuts using a table saw

9. To rip long stock, provide a means of handling it to prevent the longer end from moving around and possibly throwing you into the blade or causing an inaccurate cut.

10. Long stock can also be ripped by reversing ends.

SAFETY PROCEDURES FOR USING TABLE SAWS

Blade sharpness. Keep saw blades sharp, and select the right one for the job.

Eye and ear protection. Wear safety goggles and ear plugs while operating the saw.

Teeth and cutter direction. Be sure the saw, dado or molding head teeth and cutters point toward you as you stand on the operator's side of the saw.

Safety guard. Be sure that the safety guard is in place and ready to use.

Position. Stand to one side of the operating blade and do not reach across it. Never reach behind a saw blade to pull stock through.

Distance from blade. Do not let your hands come closer than 4 inches to the operating blade even though the guard is in position.

Stock. Stock should be surfaced and at least one edge jointed before being cut on the saw.

Saw teeth extension. When using the saw blade, set it to extend approximately ⅛ inch above the stock to be cut.

Adjustments. Make no adjustments while the blade, dado or molding head is in motion.

Freehand sawing. Always use the rip fence or the cutoff guide. Do not cut stock freehand.

Rip fence. Remove the rip fence when you crosscut.

Resawing. Resawing other setups must be carefully made and checked before the power is turned on.

Push stick. Use a push stick when you rip narrow stock.

Helpers. Other workers who help to "tail-off" the saw should not push or pull the stock but only support it. The operator must control the feed and direction of the cut.

Stopping. When the work is completed, turn off the machine and wait until the blade has stopped. Clear the saw table and place waste in a scrap pile.

Work area. Do not let small scrap cuttings accumulate around the saw table. Use a push stick to move them away. It is easy to stumble on these scraps and fall into the blade.

STAPLERS AND NAILERS

A wide variety of power staplers and nailers are available. Most of them are pneumatic (air powered), though more are becoming available as electric tools. Choice of the right tool must be based on the amount of time it saves and whether it is convenient to use on the job site. Your tool supplier can give you information on the latest developments in these labor- and time-saving devices.

Listed below are some examples of these tools and their uses.

• The pneumatic stapler drives staples that are ⅝ to 2 inches long. It is used to attach paneling, sheeting, or insulation sheets.

- The roofing stapler drives staples up to 1½ inches long and is used to staple shingles.

- The coil nailer can drive 25 different nails that are 1 to 2 inches long. It can be used for any job that requires driving nails.

- The automatic screw fastener is designed for driving screws into drywall, wood, or metal studs.

Basic Procedures for Using Staplers and Nailers

1. *Load the stapler or nailer with the appropriate fasteners.*

2. *Connect the stapler or nailer to the power source.*

3. *Check the dial or gauge for the correct setting of air or electrical impact. Too much pressure can damage the surface or drive the staple or nail through the material.*

4. *Practice on a piece of scrap stock to see whether the setting is correct.*

5. *When using one of these tools, always keep its face against the stock that is being fastened.*

SAFETY PROCEDURES FOR USING STAPLERS AND NAILERS

Manufacturer's manual. Read the manufacturer's instructions for operating these tools and follow them carefully.

Eye protection. Wear safety goggles while operating staplers or nailers.

Fasteners. Staplers and nailers, like guns, require the correct ammunition. Always use the correct size and type of fastener recommended by the manufacturer.

Power. For air-powered nailers, always use the correct pressure (seldom over 90 lbs.). Be sure the compressed air is free from dust and excessive moisture.

Direction of tool. Always keep the nose of the stapler or nailer pointed toward the work. Never aim it toward yourself or other workers.

Safety features. Check that all of the safety guards, shields, and guides are in place and working. Drive several staples, screws, or nails into a block of wood to see whether all safety devices are working as they should.

Use. During use on the job, hold the nose firmly against the surface being stapled or nailed.

Stopping. Always disconnect the power tool from the air or electrical power supply when it is not being used.

PART TWO

STANDARD CARPENTRY PROCEDURES

3
Troubleshooting Construction Problems

Construction generally follows a sequential series of steps that lead to the completion of a structure. If all goes well, the project is completed on time and without any problems. Unfortunately, it seems that problems arise throughout the construction or remodeling process. The solutions to these problems require the troubleshooting skills of the carpenter. This section deals with various troubleshooting situations that you may encounter. Many of the suggestions focus on solving remodeling problems, but the same principles can be applied to new structures.

COMPUTING THE LOAD

The amount of weight that will be placed on the foundation, floors, and roof of a structure must be carefully determined during the planning stage of construction to ensure the stability and life of the structure. Generally, the architectural plans contain the weight load that has been calculated. If you are doing work without plans and have to determine weight-load requirements, consult charts such as those on ceiling and floor joist spans included in Appendix 3. These charts give length of span, width of joists, and header requirements based on various dimensions.

In remodeling work, it is necessary to compute the additional load that will be placed on the support system of a structure if the work involves the foundation or load-bearing partition walls. It is important to compute the load requirements because the foundation may not be adequate to support the present weight, let alone handle any additional weight resulting from remodeling. To compute the loads on the various levels of the structure, complete the following steps.

1. Find the horizontal distances between bearing walls.

2. Determine what parts of the structure that the beams, headers, or joists will be supporting. (See the chart below for average weights of different parts of a house.)

Average Load per Square Foot	
Unit	**Load in Pounds**
Roof	40
Attic (low)	20
Attic (full)	30
Second floor	30
First floor	40
Wall	12

3. Multiply the load per square foot by the length and width being supported.

4. Add the loads together.

5. Using the calculated load, select the required width and thickness of support for the length being spanned.

Local building codes must be consulted to ensure safety and compliance with the municipality's support requirements.

FINDING BEARING WALLS

Before any major remodeling work is undertaken, the load-bearing walls must be located. Then shoring can be provided if these walls must be removed during the construction process. Outside walls are all load bearing to some extent. End walls in gabled single-story houses are somewhat of an exception, since these walls carry only the relatively small amount of weight that's distributed from the peak of the roof to the ceiling level. Outside walls that run parallel to the ridge of the roof carry the greatest load, since they provide most of the roof's support.

Interior load-bearing walls are not as easy to find as those outside, since it is not always as easy to see and identify the structures that are creating the load above these walls. Generally, the following interior walls are identified as load-bearing:

- a wall that runs down the middle of the length of the house,
- any wall that has a joist spliced over it,
- a wall that runs at a right angle to the overhead joists and breaks up a long span (the joist may not be spliced over the wall), and
- a wall that is directly below a parallel wall on the upper level.

Develop a sense of framing alternatives that are used in construction so that you can pick out the type used in the structure.

FLOORS

Sagging floors have a number of causes. If the home is older, the sill plate may have rotted or fallen off the foundation. Floors may sag around interior support areas if the nails have given way or if inadequate support was built into the structure during construction. Also, floors may sag if previous remodeling jobs did not provide additional weight-bearing support for added or enlarged rooms.

This frequently occurs when bathrooms are added and no provision is made for the weight of the fixtures.

If the sill plate has rotted or fallen off the foundation, it can be replaced by shoring up the edge of the house and then removing some of the siding so that the old sill can be removed and a new one installed. Then, the floor can be brought level and the joists nailed to the new sill plate. For added holding strength, metal brackets can be attached first to the joists and then to the sill plate. Threaded bolts can be run through the brackets and slowly tightened until the sill and floor are in alignment.

Interior floors that have given way can be brought back to level and secured. By placing a floor jack under the sagging area and slowly extending it, the floor will be brought back to level. Always proceed slowly during such an operation. The reason for this is that you must continually check the effects of what you are doing.

For example, when raising the floor level around a chimney, it is not unusual to have one of the floor joists catch on a brick. To proceed too quickly might crack the masonry. So work slowly and check constantly to avoid problems.

Sometimes a floor is so uneven that it must be jacked at a number of points. Perhaps the joists are so warped that they can never be brought into alignment. These problems can be corrected by removing the flooring, down to the joists, and placing 2 x 6s along the side of each joist. These 2 x 6s serve as new joists. Plane down the old joists where they protrude above the level of the 2 x 6s. A new subfloor can then be applied in preparation for the finished flooring.

A squeaking floor is a frequent problem that can be easily solved, once the cause is determined. In new construction, most subflooring is tongue-and-groove plywood that is glued and nailed to the floor joists. This method of construction virtually eliminates the possibility of squeaky floors. Therefore, this problem is generally found in older homes in which the flooring or subflooring has pulled away from the joists, allowing two pieces of wood to rub together causing the squeaky sound.

If the surface flooring is removed during remodeling, the problem can be corrected by checking for and nailing down any loose subflooring. There are also several ways of stopping the squeak without disturbing the structure. If there is access to an area where the bottom of the subflooring and the joist are exposed, slip a wedge between the top of the joist and the bottom of the subfloor. The wedge will stop any movement and eliminate the squeak. Another way of stopping squeaks from the underside of the floor is to run a screw through the subfloor into the finish hardwood floor and tighten the screw until the subfloor and finish floor are securely resting on each other. If you cannot gain access to the floor joists and subfloor, you can drive a spiral flooring nail through the carpet or wood floor, subfloor, and into the joist. Two nails should be driven into each finished board and angled toward each other. If you are nailing into hardwood, drill a pilot hole first. Use a nailset to drive the nail below the surface where it will not be seen or felt.

If a floor is uneven due to cracks or openings between the subfloor pieces, it is best to put underlayment down before laying the finish floor. The underlayment will provide a smooth surface upon which the finish floor can be laid.

FOUNDATIONS

Before you begin a remodeling project, inspect the foundation for stability. Older foundations are generally made from poured concrete or stone work. If the foundation is made of poured concrete that is crumbling and moving, several methods can be used to stabilize it. One method is to pour new walls of concrete that are 6 to 8 inches thick around the interior of the basement or crawl space. This method does not require excavation around the exterior of the structure, but it does reduce the size of the interior area by the thickness of the newly poured walls. Reinforcement bars should be inserted into the old walls so that ties are created between the old and new walls.

Another method of stabilizing a concrete foundation is to pour new walls against the outside of the foundation. This method requires excavation of the foundation area.

If the condition of the foundation is such that neither of these methods will be effective, the structure can be lifted up and a new foundation can be built under it. Once the building has been lifted the construction procedure is the same as that used in constructing a new basement or foundation.

Structures that have stone or rock and mortar foundations that require stabilizing can be made secure by using the procedures just discussed, or the building can be set on beams while still resting on the rock foundation. If beams are used, the first step is to calculate the weight load of the structure. The beams (generally steel) are then positioned so that the load will be evenly distributed. A concrete pad is poured wherever a house jack will be placed. These concrete pads become footings for the house. They must be thick enough to carry the weight placed on them without moving. They are usually 12 to 18 inches thick and 12 inches square. Local building codes will give you the exact requirements.

The beams carry the weight of the structure. The foundation walls remain in place and continue to protect the structure from the elements. The house, in effect, is floating on the foundation but will not continue to settle or fall away from it.

Basement floors in older structures are often cracked or broken, creating an uneven surface. If a finished floor is going to be installed, the concrete floor can be leveled and made ready for finished flooring by patching the uneven areas with one part mortar cement and three parts fine sand. If portions of the floor have broken up, remove the pieces that stick up and clean out any debris in the low spots. The depressions in the floor should be roughed with a hammer and cold chisel so that the new concrete will have a surface to which it can adhere. Fill in the depressions and then work the concrete to level, just as you would any other concrete project.

If the basement has a water problem, it may be attributable to the lack of drain tile along the outside of the foundation. If the remodeling project is a major one and a dry basement is desired, then drain tile and gravel should be installed for proper ground drainage.

Other problems, such as improper grading, may also cause basement water problems. Regrading or burming the property may be necessary.

OLDER STRUCTURES

Before starting a remodeling project, walk through the area to determine what must be done to create the desired effect. New construction follows a set of plans that cover every step involved in completing the project. Remodeling involves visualizing a finished project while viewing an existing structure. When walking through a structure, you must determine the construction techniques that were used. Houses built prior to the 1930s probably will have balloon framing. Balloon framing's main feature is its long studs that run from the sill on the first floor all the way to the plate on which the rafters rest. These are called building height studs. They may be spaced anywhere from 12 to 24 inches on center (O.C.). It seems that in older construction no standard was used, so you never know what you will find when you expose the studs. Also, if the home is old, it probably has been remodeled several times before. This means that old doorways, windows, and walls may have been changed or removed. As a result, the studs can be anywhere.

A common problem that you may encounter with balloon framing is a wall that has moved out of alignment with the structure. A way of removing the resulting bulge in the wall is to remove the siding and interior wall covering, shore up the floor and ceiling joists, and then cut the studs that have bulged out. The cut studs then can be brought back into alignment. Nail a scab stud along each cut of the studs and nail the studs to the sill. The studs should then be secure and ready for sheathing.

The ceilings in houses built before the 1930s were traditionally 9 or more feet in height. If such a house is being fitted for plumbing and central heating, the difference in space between the standard 8-foot ceiling height and the house's current ceiling height can be used for running the lines. Before you do any structural work, determine the location of current lines and decide where they must be placed to avoid cutting into them. When cutting into walls or ceilings, shut off all power to the area; as an extra precaution, use double insulated tools with plastic or rubber sleeves.

Special problems arise when working on older homes, especially when trying to match two different types of construction and lumber dimensions. Older lumber was cut to the actual measurement that was specified. Modern lumber is based on a specified measurement but is not cut to it. For example, older 2 x 4s are actually 2 inches by 4 inches. Modern 2 x 4s measure 1½ inches by 3½ inches. When bringing old and new walls together, these differences must be considered. You should check three areas when combining old and new materials and construction.

1. *Foundations.* Adjust the level of footings to accommodate the difference between old and new foundation units. You need to visualize at what level the floor will be finished to ensure a proper match of the finished units.

2. *Ceiling height.* Older homes generally differ in ceiling height as a result of nonstandard building. When changing ceiling height make sure that the proposed ceiling height meets local building codes and has space for utility lines between the old and new ceilings.

3. *Roof lines.* When you are bringing together two roof lines, adjustments will have to be made to get the desired match. If the match cannot be made as desired, alter the plan until you are satisfied.

ROOFS

Since roofs are constantly exposed to the elements, they must be inspected on a regular basis in order to prevent problems or excessive damage. By climbing up and inspecting the roof, it will be evident whether any shingles are missing or wearing out. Asphalt shingles that are worn or missing most of the granular surface or curled up on the edges must be replaced. Wooden shingles that are split or missing must be replaced. By making some general observations, you can prevent serious problems.

If the roof has been neglected for a number of years and a new roof is being installed, be sure to inspect the eaves, rafter ends, and soffits. Water may have leaked through the shingles or collected in the gutters, covering the fascia boards and rafter ends for long periods of time. If the rafter ends are rotted, the roofing and sheathing must be removed and new rafter ends installed. The rafter ends can be repaired by splicing new rafter ends onto the existing rafters and then placing a scab on the side of each splice for additional support. If you prefer not to splice new ends on the rafters, you can just scab along side the existing rafters and cut the ends off to match. Sheathing that has any evidence of rot should be replaced, since it will continue to lose its supporting ability as it ages.

If the roof leaks, you must try to trace the leak's source if you are not going to replace the whole roof. Tracing a leak can be a frustrating task, since often water escapes through the interior walls of a structure at a different point from which it enters through the roof. To find a leak of this nature, you will have to climb up into the attic during a storm, see where the water is coming in, and mark the spot by driving a nail from the underside of the roof through to the outside. Instead of waiting for a storm, you could have someone spray a gentle flow of water on the roof while you watch for the leak. When you are ready to fix the leak, the nail sticking up through the roof is your guide to the leak point. You can fix the leak by laying new shingles or by patching with roofing cement.

If the attic has a finished ceiling, the job of finding the leak is more difficult. The best way to find the leak in this case is to carefully measure from the leak point to the end of a wall or some other reference and transfer the measurement to the outside of the roof. You can do this by measuring the number of joists or rafters from the wet spot to the chimney, end of the house, or some other well-defined area. Multiply this number by 16 inches, divide by 12, and this will give you the number of feet that must be measured to get to the leak area.

SCAFFOLDS

Scaffolds are used in areas where access is difficult, and provide a secure, rigid platform on which to stand and work. Scaffolds are set at a height that allows the worker to do the work without excessive stooping and reaching.

Scaffolds can be constructed on the job site by using prime

straight 2 x 4s as uprights. The uprights should be placed on planks to prevent them from sinking into the ground. The cross ledgers are made of 2 x 6s that are about 4 feet long. They are nailed with 16d nails to the uprights using three nails per joint. The braces can be made from 1 x 6 lumber and are fastened to the uprights with 10d nails. The platform is made of 2 x 10 planks that are free from large knots and that are not split or broken. The planks should be spiked to the cross ledgers to prevent slipping.

Most scaffolding is now manufactured and can be reused. The scaffolding frames are made from steel or aluminum and come in sections. The frames range in size from 2 to 5 feet wide and from 3 to 10 feet high. The basic units are set up and joined vertically and horizontally. This type of scaffolding can be purchased or rented as needed.

There are other forms of scaffolding, such as roofing brackets and ladder jacks, that allow you to work at various heights without having to hang from a ladder. All of these scaffolding devices must be set up properly for safe use. Each year, many workers are hurt or killed as a result of improper placement or use of scaffolding.

SHORING

When doing any type of construction work, support must be added to prevent the structure from sagging or collapsing when removing load-bearing walls or parts of load-bearing walls.

The process of adding support to the structure is called shoring. A simple method is to create a T support by nailing two 2 x 4s together in the shape of a T. A plate is set below and above the T. The T is then driven into place securely. Another method of shoring is to create a short wall of 2 x 4s. The wall is made so that it can be set in place securely; if necessary, wedges can be driven between the bottom plate and the floor to make sure that the wall does not move. Alternatively, jack posts can be placed on plates (4 x 6s) and then adjusted for a secure fit. Remember, if there is any doubt as to whether the structure has adequate support, take extra precaution and use shoring.

SIDING

Siding problems can detract from the appearance of the structure. The problems may be caused by the method of installation or a defective surface covering. The siding must be securely fastened to the wall studs. If a stud has been missed, the siding can curl or pull away from the sheathing. This problem can be easily fixed by locating the missed stud and nailing the siding to it.

If paint or stain on siding is peeling, flaking, or chalking, the problem may be the result of improper application or the lack of ventilation behind the exterior surface. If the method of application is the source of the problem, then the surface must be scraped to remove the loose pieces. Sometimes, it's possible to remove flaking paint with a water-pressure sprayer. The sprayer can be rented, and it provides a less tedious and less time-consuming alternative to scraping. If a ventilation problem exists, small vents should be installed at the bottom and top of each stud section that lies beneath the surface where flaking or peeling is evident. The details for installing these vents are discussed in the section on insulation.

STAIRS

Common problems that occur with stairs in older homes are loose or missing treads and risers, squeaky treads, and loose stringers. It is obvious when treads and risers are in need of repair or replacement. Squeaky treads are handled in much the same way as squeaky floors, by using nails or wedges from the bottom or by driving nails down through the treads into the stringers. Another way of fixing squeaky stairs is to fasten metal shelf brackets on the underside of the tread and the back of the riser. This will hold the tread securely to the riser and eliminate any movement that can cause a squeak.

Loose stringers can be corrected by driving the stringer back into position with wedges located between the wall and the stringer. If the gap between the stringer and the step is more than ½ inch, do not use wedges because you will probably split the stringer. It is better to replace all of the steps to ensure a solid stairway.

WINDOWS AND DOORS

A problem that can come up with any construction project is work that is out of square or plumb. New windows and doors, for example, must be checked for level and square when first set into the structure. In older structures, the problem will be to readjust for square, particularly for doors.

Doors that do not fit squarely can be corrected by shimming out the hinges with cardboard until the desired fit is accomplished. Another way to square doors is to remove stock from the edge that is sticking. This is done on a trial-fit basis to avoid removing too much stock and creating a gap. Also, wedges can be inserted where needed to adjust the level of the floor or the square of the casing.

Troubleshooting square is often a matter of common sense and may be achieved by simply stepping back and looking at what is causing the area or item to be out of square. A frequent reaction is to drive the item into square using force. This generally does not work, and if it does, it is usually only a temporary solution.

Windows and doors generally are replaced for either of two reasons: (1) to remove rotted defective materials or (2) to improve the style of the structure. If a rotted window or door is being replaced, check the area around the opening and remove all of the rotted material. If a water leak existed around a window for many years, chances are that the studs, headers, and rough sill must be replaced. Setting a new window or door into an opening that is not secure will solve the problem for only a short time. If you are putting in a new window or door to enhance the appearance of a structure, the job generally involves cutting the rough opening to size, roughing it in with the stud work, and then setting in the window or door. Manufacturers provide instructions for installation, since the windows or doors generally come to the job site in kit form.

4
The ABCs of the Trade

CABINETMAKING

Cabinets serve two functions in the home or commercial setting. The most important is storage. Storage must be adequate to hold all the items needed in a kitchen, closet, or bathroom. Cabinets must be constructed so that they can accommodate various sizes and shapes of household appliances, dish shapes, and foodstuffs. Cabinet space must also be flexible in order to accommodate changing storage needs. Commercial cabinets are frequently specially designed for the products they will display or hold. The methods and materials used in constructing these cabinets differ from those for home cabinets.

The second major function of cabinets is to enhance the area in which they are located. In some cases, cabinets complement a specific type or style of furniture, wall decoration, or household theme. Commercial cabinets can serve in this capacity as well, but most often their exterior components are selected for the ability to withstand wear. Final selection of any exterior cabinet component depends on the customer's assessment of location, wear, decorating theme, and price range. Make sure you explore all of the options available in countertops, drawers, shelves, doors, and exterior finishes. These options can be found in brochures or other illustrative materials from the various cabinetmaking firms.

Cabinets are produced in three ways:

1. They can be built on the construction site.
2. They can be built at a mill and shipped to the construction site in a knocked-down form.
3. They can be mass-produced in a factory and delivered assembled and ready to set into final position.

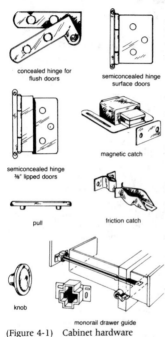

concealed hinge for
flush doors

semiconcealed hinge
surface doors

semiconcealed hinge
⅜" lipped doors

magnetic catch

pull

friction catch

knob

monorail drawer guide

(Figure 4-1) Cabinet hardware

Cabinets built on the construction site are a rarity today. Even with custom cabinets, the work will normally be done in a mill, and the cabinets will be shipped to the site. Once the cabinets are delivered to the site, they will be installed according to a predetermined plan. The job of installing cabinets requires skill and carefull attention to detail, since this is one of the final finishing tasks.

CABINET HARDWARE

Cabinet hardware is installed as a finishing touch to the cabinets. The knobs, pulls, hinges, catches, and other metal fittings that are put on cabinets come in different designs, forms, and finishes. (See Figure 4-1.) By consulting manufacturers' brochures, you can become familiar with the various hardware options that are available. Care must be taken in selecting an appropriate size, style, material, and finish. Purchase hardware prior to constructing doors and drawers, since it often dictates size and details of work.

Basic Procedures for Installing Cabinet Hardware

1. *Lay out hardware on cabinets. Drawer pulls usually look best when they are located slightly above the center line of the drawer front. They must be level and centered horizontally.*

2. *Hinges for cabinet doors should be located 2½ inches from the top and bottom edges of each door. This location gives the maximum*

support to the door while keeping the hardware away from the edges where the screws might split the wood or tear away from the door. (See Figure 4-2.)

3. Door pulls are convenient when located on the bottom third of wall cabinets and the top third of base cabinets.

4. Swinging doors require some type of catch to keep them closed. There are a number of these devices available. One factor in selecting a type of catch is the noise that it makes. Some catches are quite noisy in operation, while others are almost noiseless. The catch should be located as near as possible to the door pull.

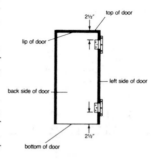

(Figure 4-2) Hinge placement

5. Cabinet hardware on the outside of the cabinet should be prefitted before the final finish is applied.

6. When drilling the holes for the hardware, use sharp drill bits to prevent splitting and splintering. If possible, use drilling jigs when installing hardware to ensure accuracy and reduce the chances of marring the surface.

CABINET MATERIALS

Cabinets are made from a wide variety of materials, and the final choice of material is based on personal preference and cost. Low-priced cabinets are usually made from panels of particle board with a vinyl film applied to exposed surfaces. The vinyl can be printed with a wood grain pattern that has the appearance of genuine wood, or it can be a solid color.

High quality cabinets are made from veneers and solid hardwoods, including such species as oak, birch, ash, and hickory.

Inside parts may be made from hardboard, particleboard, or waferboard, depending on the quality of the cabinet and the amount of stress put on the material. Within the cabinets, certain metal or plastic items may be installed, such as revolving shelves, dividers in drawers, or racks for storing pan lids.

Specialty items, such as slide-out breadboards, shelves, cutting boards, or canned goods storage units, may also be installed. These items may require the use of several different materials in combination with each other. Most of the special inserts come as whole units from the manufacturer and only require installation. Others (such as the plastic cutting board) may require that you rout out a space in the countertop.

CABINETS AND BUILT-INS FOR OTHER ROOMS

Cabinets used for rooms other than the kitchen come in a similarly wide variety of styles. They are either custom built or factory made. Most often, custom-built cabinets are preferred for room dividers, storage cabinets or built-in bookcases, since it would be difficult to find factory-built units that suit such individualized uses and space requirements. The linen closet is a common built-in storage unit that a carpenter is called upon to build. This job requires the use of basic cabinetmaking skills but the work space is often very cramped and requires close fits that are hard to reach for nailing.

CABINET SIZES

The blueprints for a job include the location and design of cabinets for the building. A specifications sheet is normally included with the plans. This sheet outlines the cabinets to be used, any special joints, space fillers, materials, or modifications that need to be made to the site or cabinets.

Typical kitchen cabinets include base and wall cabinets. The base cabinet is usually 36 inches high and 24 inches deep, including the countertop. Most wall cabinets are 30 inches high and 12 inches deep and contain two movable shelves. The distance between the base cabinet and the wall cabinet is usually 18 inches. (See Figure 4-3.)

Keep in mind that when any cabinets are installed, the customer must be remembered. For many people, the 36-inch height is too high; a countertop 32 inches high might be much more comfortable to use. These modifications should be noted on the blueprints; if they are not, inquire about any changes that might better suit the customer's needs.

COUNTERTOPS

The selection of the appropriate material, color, and texture for a countertop is very important, since the top will be seen and used all the time. The most commonly used material for countertops is plastic laminate. This material is usually 1/16 inch thick and offers resistance to wear, harsh household chemicals, grease, oil, and boiling water. This material, although

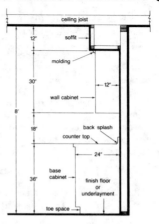

(Figure 4-3)   Cabinet size

very tough on the surface, is very fragile unless it is bonded onto a solid base or some form of core material. The two most commonly used materials are smooth-finish plywood or particleboard. The plywood must be very smooth or the grain of the wood may start showing through the laminate, causing the laminate to wear through at the points where the grain has caused it to lift. Particleboard is generally preferred because it is very dense, does not have grain, and is less expensive than plywood.

There are two common forms of countertops used in construction: rounded-lip edge and square edge. (See Figure 4-4.) The rounded-lip edge counter-

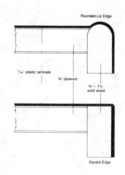

(Figure 4-4)   Countertop types

top is normally made at the manufacturing plant and shipped to the work site. The reason for this is that it takes special equipment to heat and form the plastic laminate. The rounded-lip edge countertop not only has the slightly raised lip at the front of the counter edge, but has a rounded back-splash guard that curves up to a height of 5 or 6 inches. This countertop comes as one-piece—core material and laminate bonded together—and is ordered by the foot.

The square-edge countertop may be made in a plant and ordered for the job by the foot just as the rounded-lip edge countertop. Since no special bending is required in applying the laminate, the square edge countertop can be constructed on the job site. This is an important feature if a counter is needed that is not standard size.

Basic Procedures for Working with Laminates and Installing Countertops

1. *Plastic laminates are sold in sheets. Before ordering the laminate, carefully lay out the exact length and width needed.*

2. *Laminates can be cut to rough size with a handsaw, table saw, portable saw, or router. Use a fine-tooth blade and support the material close to the cutting edge.*

3. *Cut the laminate to rough size approximately ¼ to ⅛ inch larger than the desired finished work.*

4. *Coat the backing material (particleboard, plywood) with contact cement. To spread the cement, use a brush or metal spreader. On larger surfaces, it is easier and quicker to use the metal spreader. Always work in a well-ventilated area. Shut off all pilot lights, and read product instructions and precautions when using contact cement.*

5. *Coat the laminate back with the same contact cement. Use a brush or a metal spreader. All of the surface must be completely covered with a glossy film.*

6. *After the cement has been applied to both surfaces, let them dry. You can check to see whether the cement has dried enough by pressing a piece of paper against the cement. If no cement sticks to the paper, then the surfaces have dried enough to be assembled.*

7. *Bring the two pieces into exact position. There is no second chance. On large pieces, a sheet of heavy wrapping paper can be put between*

the two pieces while they are put in position. This paper is called a slip sheet. If you are doing a lot of laminate work, the slats from venetian blinds work very well. They are thin, flexible, and easy to slide out as needed. If a slip sheet of any kind is used, check the position of the two pieces and then move the slip sheet out until one edge can be bonded; then remove the entire sheet and apply pressure.

8. Total bond is achieved by applying momentary pressure. To finalize the bonding, a roller can be used over the laminate. In corners where the roller can't reach, use a block of wood and tap it with a rubber mallet.

9. To trim the laminate to exact size, use a router with a laminate trim cutter or a laminate trimmer (an undersized router with a small base for good router control on edges). These tools remove the excess material and leave a smooth edge. Wear eye protection when trimming laminate with a router.

10. If an angle cut is needed, plan this out very carefully—a mistake can ruin the entire countertop by making it too short after the cut is made. Angle cuts can be made on a radial arm or table saw. Use a hollow ground combination blade to get a smooth, even cut. Make sure you have adequate help to steady the countertop while cutting. The slightest movement can throw off the degree of cut and cause a poor fit when the counter is assembled.

11. Sink cutouts or other openings can be made with an electric saber saw.

12. Apply and trim the edges of the countertop in the same manner as the work surface.

13. Bevel the corners and edges of the laminate. The bevel makes the counter more durable by eliminating sharp edges that might break off. Removing sharp edges also avoids the possibility of injury while working at the counter. The bevel edge can be created by using a laminate bit in the router or by making downward, angled strokes with a mill file. Number 400 wet-or-dry abrasive paper can be used for final smoothing.

14. Set the countertop in place.

15. Fasten the countertop to the base cabinet by predrilling and screwing through the front corner blocks into the top.

Installing Rounded-Lip-Edge Countertops

1. *Ends are needed for the rounded-lip-edge countertop. Carefully determine the number of left and right ends you will need. On most jobs, you will need only one set of left and right ends.*

2. *Fasten the rounded-lip edges on the ends of the countertop with contact cement. Some edges are cut slightly oversized, and you will need to rout the edges to fit the countertop exactly. Use the same procedure as described in the section on working with laminates.*

3. *Remove any excess cement from the counter by wiping the laminate with a cement thinner.*

Installing Square-Edged Countertops

1. *Square-edged countertops can be purchased ready-made, as can the rounded-lip-edged countertops. If you use a pre-made countertop, the assembly is the same as the rounded-lip-edged countertop.*

2. *If the counter is being made on site, the base material must be cut to size, and then the counter edge is applied to it. The edge material is normally ¾-inch solid wood and is glued flush to the base material. This increases the width of the counter by ¾ inch and must be considered when measurements are taken.*

3. *Apply the plastic laminate using the method described in the section on working with laminates.*

Cutting Sink Openings

1. *Lay out the exact location of the opening, and mark the laminate with a felt-tip pen. This marking method does not mar the surface, and excess marks are easily removed.*

2. *Drill a pilot hole of approximately ⅜ or ½ inch in size. The hole must be large enough to allow the blade of the saber saw to enter. When drilling pilot holes, make sure you stay inside of the line as this is scrap stock.*

3. *Place the blade of the saber saw in the pilot hole and proceed to cut along the marked lines. Most sink cutouts are rounded on the corners or round in total shape so there is no problem in following the layout lines. If the cutout is square, pilot holes must be drilled in all four corners.*

4. *Complete the cutting procedure and remove the cutout.*

5. *Smooth any rough edges on the cutout of the countertop.*

6. *Check for sink fit.*

FACTORS TO CONSIDER

- Use contact cement in a well-ventilated area.

- Remember that when the two workpieces touch each other, the bond is permanent, so make sure both pieces are properly positioned before you bring them into contact.

- Laminates have a very durable surface but are very fragile when worked on without backing material. They break easily at points where irregular cuts have been made. Be careful when moving tools around laminates to avoid scratching the finished surface.

DOORS

Doors on cabinets serve two purposes: they (1) close off storage space and (2) provide decoration and finish to cabinets. The primary concern is over the latter, since almost any material could be used to block off the storage space. The door designs selected must complement the cabinets and fit into the decor of the structure. The grain and color of the door fronts should match the drawer fronts and face frame.

Cabinet doors come in four types and are named according to the method used to fasten them to the face frame or the way they move to close off the storage space. The different types are described below.

Flush door. The flush door fits into the opening and does not project outward beyond the face frame. It is made about $\frac{1}{16}$ inch smaller than the opening on each side so it will open and close without rubbing.

Flush-overlay door. This door has square edges like the flush door, but it is mounted on the outside of the frame. The edges either partly or wholly conceal the face frame.

Offset or lip door. The offset door is rabbeted along all edges so that part of the door is inside of the door frame. A lip extends over the frame on all sides, concealing the opening. The offset door is easy to fit; its edges cover part of the face frame so no cracks can show at the edges. The outer edge of the door is usually rounded or decorated by another design or edge treatment.

Sliding door. The sliding door moves in a groove that is cut in the face frame or on plastic or metal rollers set in a track. A sliding door is used in areas in which a hinged door would take up too much space by opening into the room.

Like drawer fronts, cabinet doors come in a wide variety of prefabricated styles, shapes, and designs. Prefabricated doors can be delivered cut to exact dimensions with lipped edges already machined (if specified).

A door is constructed as a solid panel or a combination of frame and panel. The solid-panel door is usually made from ¾-inch plywood. The frame-panel door is made of solid wood frame. The frame requires four or more pieces. There are two vertical side pieces called stiles and two horizontal pieces (a top and bottom) called rails. Some doors have a middle horizontal piece called a cross rail.

For most cabinet doors, the thickness of the frame is ¾ inch, but the frame for an ornate door can be up to 1 inch thick. The center panel in frame-panel doors may be constructed from ¼-inch plywood or solid wood. A slot is cut around the frame into which the panel is set. The strength of the door depends on the method used in joining the corners of the frame. The mortise and tenon joint is most often used in quality cabinet doors.

The frame of the cabinet door will hold the plywood and solid wood in place so there will not be any warping. Solid wood is not used in making the entire cabinet door, it has a tendency to warp when cut to standard cabinet size.

Basic Procedures for Installing Cabinet Doors

Lip and Flush-Overlay Doors

1. *Cut the doors to the size of the openings plus the amount required for the lip all the way around the door. Allow clearance of about ¹⁄₁₆ inch on each edge. Make an additional allowance for the hinges that are not recessed in the door or frame.*

2. *Form the lip by cutting a rabbet along the edge. This can be done with a table saw or a shaper.*

3. *Check the fit of each door in its proper opening. Once the correct fit has been achieved, mark the door and the opening for ease of final assembly.*

4. *Remove all machine marks, and sand the surface in preparation for finishing.*

5. *Finish doors.*

6. *Put on hinges. The lip door requires a special offset hinge. Several designs are available, and choosing one is a matter of personal preference.*

7. *Attach doors to face frame.*

Flush Doors

1. *Select hinges. Normally flush doors are installed with butt hinges; however, surface hinges, wraparound hinges, knife hinges, or various semiconcealed hinges can be used. Select a hinge that can support the weight of the door onto which it will be fastened.*

2. *Cut the door to fit the opening. There should be $1/16$-inch clearance on each edge. The total gain (space) required for the hinge can be cut entirely in the door. For really fine cabinets, the gain should be taken equally from the door and stile.*

3. *Mount the hinges on the door.*

4. *Mount the door into the opening and then set the hinges into the frame. When first mounting the door, set only one screw into each hinge until the fit desired is achieved; then set the rest of the screws.*

5. *Put stops on the door frame so the door will be flush with the surface of the opening when closed.*

Sliding Doors

1. *Cut the doors to fit the opening. The final dimensions of the doors will be determined by the depth of the grooves into which the doors will fit and the amount of overlap needed where the doors come together. Normally, a 1-inch overlap is allowed, although the amount of overlap may vary depending on the kind of installation.*

2. *For finer sliding doors, a rabbet is cut around the doors. The rabbet should have a $1/16$-inch clearance in the groove. The top rabbet should be cut deep enough so that the door can be inserted and removed by raising the door into the extra space.*

3. *The grooves in the door frame should be $1/16$ inch wider than the doors for easy movement.*

4. *For heavier doors, installing a metal or plastic roller track will provide easier operation.*

5. *Finish doors.*

6. *Set doors into place. If there is resistance in sliding the doors, apply a coat of wax.*

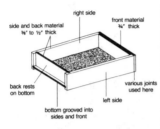

(Figure 4-5)   Parts of a drawer

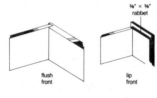

(Figure 4-6)   Types of drawer front

DRAWERS

Drawers must be constructed from durable materials because they are used all the time and in many cases have heavy items placed in them. Drawers are normally constructed after the cabinets are made to ensure that they will fit the specific cabinets in which they will be used. (See Figure 4-5.)

There are two types of drawers: lip and flush. (See Figure 4-6.) Flush drawers are commonly used in furniture and must be carefully fitted. Lip drawers have a rabbet along the top, sides, and bottom of the front. This style overlaps the opening and is much easier to construct.

Drawers have five parts: front, right side, left side, back, and bottom. Drawer fronts are made of either solid wood or veneered plywood. They are usually not less than ¾ inch thick. Stock for drawer sides is either solid wood or plywood. It is usually ⅜ or ½ inch thick. The bottom is made from ¼-inch plywood. The back is made from the same stock as the sides.

High quality drawer sides are made from either oak or maple. Inexpensive cabinets may be made of pine, poplar, or willow. Different kinds of joints are used for attaching the sides to the front. The most difficult joint to make is the dovetail, due

(Figure 4-7)   Dovetail joint

to the complicated setup required to cut the joints. (See Figure 4-7.) The drawer sides have grooves to receive the bottom piece.

Joints for fastening the drawer parts together are chosen based on the quality of drawer desired. It takes more time to produce good joints; therefore they are more expensive. (See Figure 4-8.) Regardless of the style of joints selected, the joints between the drawer front and the drawer sides must be strong, since these areas are the most common breakage points.

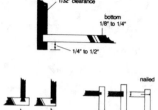

(Figure 4-8) Common drawer joints

Drawers can fit snugly in the opening without any side-mounted hardware needed for smooth operation, or they can be made smaller than the opening to accommodate side-mounted metal drawer glides. If you decide to use metal drawer glides, purchase the hardware before constructing drawers and read the manufacturer's specifications for necessary clearances (typically about ½-inch clearance on each side).

Basic Procedures in Drawer Construction

1. *Select the material for the drawer fronts. Try to match the grain patterns so that the fronts will blend with each other if several drawers are situated side-by-side. Prefabricated drawer fronts are available for standard size drawers. They come in a variety of finishes and can cut down on construction time and cost.*

2. *Cut the drawer front to the size of the opening. If the drawer is to have a flush fit, the clearance should be ¹/₁₆ inch all around. For lip drawers, the width of the rabbet all the way around the drawer front must be added to the dimensions.*

3. *Select and prepare the stock for the sides, back, and bottom.*

4. *Cut grooves for the bottom in the front and sides.*

5. *Cut joints in the drawer front that will hold the drawer sides. For drawer fronts with a lip, cut the rabbet first, then the joint.*

6. *Cut the matching joint in the drawer sides. Be sure to cut a left and right side for each drawer.*

7. *Cut the required joints for the drawer sides and back.*

8. *Trim the bottom to the correct size. Make a trial assembly. Make any adjustments needed to ensure proper fit.*

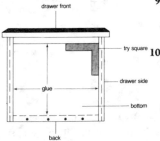

9. *Disassemble and sand all parts. The top edges of the sides can be rounded at this time.*

10. *Make the final assembly. Be sure to get glue into all of the grooves and joints. There are no specific steps to follow in terms of sequence. Proceed in an order that gives you time enough to get the drawer assembled before the glue sets.*

(Figure 4-9)   Making a drawer

11. *Carefully check each drawer for squareness. Once the drawer is square, drive two or three finishing nails through the bottom into the back. The nails along with the glue will hold the drawer square. Wipe off excess glue. (See Figure 4-9.)*

12. *After the glue has cured, fit and adjust each drawer to a specific opening. When a drawer fits, mark the drawer bottom and the inside frame of the cabinet. This will eliminate guesswork later during final assembly.*

FACTORS TO CONSIDER

■ Take care in selecting drawer fronts. They are always visible reminders of the work that was done.

■ Set up the saws with fixtures so that you can repeat the cuts on all of the drawers without having to reset the saws. This will save time and ensure accuracy.

■ Be sure to wipe off excess glue while the glue is still wet. Dried glue creates a barrier that will not allow stain or finish to penetrate the wood.

FACING

Facing strips are applied to the frame of the cabinets for both beauty and structural support. In manufactured or shop-built cabinets, these strips are often assembled into a framework (called a face frame) before they are attached to the basic cabinet structures.

As with the door-frame parts, the vertical members are called stiles and the horizontal members are called rails. If you are installing preassembled units, you will not be working with the face frame unless you are attaching doors to it. If so, proceed with caution, because in most cases the face frame will be finished and the surface can be easily marred. In selecting the wood to be used for the strips, choose hardwood of the same kind as that used in the drawer fronts and doors. Match the grain and wood color as you have done on all other parts of the cabinets.

Basic Procedures for Installing Facing Strips

1. *Before cutting the strips to size, hold each strip in place where it will be used and mark it. This will allow for variances in the cabinet frame and give a better fit.*

2. *Install the stiles first and then the rails. Sometimes, an end stile is attached plumb and rails are installed to determine the position of the next stile.*

3. *Glue and nail the parts into place. When nailing hardwoods, drill nail holes where splitting is likely to occur.*

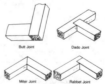

4. *There are a number of joints that can be used to fasten the face frame. The most commonly used are the butt joint* (Figure 4-10)   Cabinet frame joints *and the dado-rabbet joint. The dado is cut in the stile, and the rabbet is cut to fit in the rail. On finer custom cabinets, dowels are used to make the joint even stronger. (See Figure 4-10.)*

FINISHING PROCEDURES

Cabinets accent the room in which they are located. Care must be taken to ensure that the final finish is applied correctly and that it brings out the beauty of the wood.

Care must be taken in installing finished factory-built or preassembled cabinets on the job site. Make sure that the face sides of the cabinets are well protected to prevent tool tracks when setting the cabinets in and installing cabinet hardware.

FRAMING

Framing is the point in the construction process at which all of the parts start coming together. Accuracy is tested here, because as the assembly proceeds it will become clear whether care has been taken in the cutting and fitting process.

Basic Procedures for Framing in Cabinets

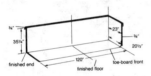

(Figure 4-11)   Framing-in a cabinet

1. *The basic layout can be made directly on the floor or wall surface. Each end panel and partition is represented with two lines. (See Figure 4-11.) Be sure these lines are plumb, since they can be used to line up the panels when they are installed.*

2. *Construct the base. The base is made of straight 2 × 4s nailed to the floor and to strips attached to the wall. (See Figure 4-12.)*

3. *Leveling the base. If the floor is not level place shims under the various members of the base until level is achieved.*

4. *Where the 2 × 4s are exposed, cover them with finished materials. On the front edge, use base molding.*

5. *Once the base is completed, cut and install the end panels. Attach a strip along the wall between the end panels, and make sure it's level. Nail it securely to the wall studs.*

6. *Cut the bottom panels and install them in place on the base.*

7. *Cut and install the partitions. They are notched at the back corner of the top edge so that they will fit over the wall strip.*

8. *Plumb the front edge of the partitions and end panels. Secure them with temporary strips nailed along the top.*

9. *Use the same basic procedure when constructing wall units. Make layout lines directly on the wall and ceiling. Attach mounting strips by nailing*

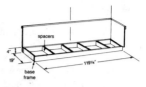

(Figure 4-12)   Base of a cabinet

through the wall surface into the studs. At inside corners, end panels can be attached directly to the wall.

10. *During the basic framing operations, care must be taken to allow enough clearance for built-in appliances. Rough drawings are furnished by the manufacturers of the appliance units.*

11. *After the frame or panels have been assembled, facing strips are attached.*

FACTORS TO CONSIDER

- Check and recheck the layout before proceeding. An error at this point can cost much in time and effort.
- Often, during the assembly of the frame, a better fit can be achieved by holding the parts in place and then marking for fit and layout. This will accommodate slight irregularities in the layout, cutting, and assembly process.

INSTALLING PREBUILT CABINET UNITS

Almost all cabinets used today are made in factories that specialize in cabinet construction. These factories can mass-produce cabinets that are attractive, well built, and less expensive than those that are built on site. Factory-built cabinets will come to the job site in one of three forms.

1. Disassembled. The disassembled or "knocked-down" cabinets consist of parts that are cut to size and ready to be assembled on the job.

2. Assembled but not finished. The assembled but unfinished cabinet is ready to set in place. Hardware is included but not installed. All surfaces are sanded and ready for finish. After installation, finishing materials and procedures can be coordinated with doors and inside trim, ensuring a match of finish work throughout the room and/or house.

3. Assembled and finished. The most commonly used cabinets are those that are assembled and finished. Manufacturers offer a variety of shades and colors of finish that will match almost any finish being applied on the job. Also, since the finish is applied in a carefully controlled setting, it is applied evenly without major running or other finishing defects. The finishes used are highly resistant to moisture, acids, and abrasions. These finishes cannot be applied on the work site without special equipment.

Three options are available in using manufactured cabinets. First, the entire cabinet assembly can be made up of factory units. This option requires that you make the space that you are filling with cabinets fit the dimensions of the factory units. Second, factory units can be used in combination with custom-made cabinets. Third, factory units can be ordered to fit the specific job and shipped to the job site. This option allows you to order the cabinets to fit the dimensions of the job.

Basic Procedures for Installing Prebuilt Cabinet Units

1. *Decide whether you are going to install the base cabinets first or the wall units. Procedures are given for beginning with either set.*

2. *Hanging wall units first. When the layout has been completed and the studs have been located, the wall units are lifted into position. They are held in position with padded T-braces or sawhorses that have been built to the correct height. A check is made for fit and level. If any trimming is required, the cabinet is removed, the trimming is completed, and the cabinet is returned to the wall, resting on the support.*

3. *Fasten the unit into place with nails or screws driven through the fastening strip provided at the back of the cabinet. Make sure you have hit the studs because the drywall will not support the unit when it is loaded with dishes or pans.*

4. *Setting the base units first. The only difference in procedure is that*

the base units, once installed, can be used as supports while setting the wall units into position.

5. *The base cabinet units must be installed level in both directions and have a tight fit against the wall. If the cabinet is not level, it is adjusted to level with wood shims. The unit is then screwed to the wall. If the base cabinets cannot be lined up to be fastened to studs behind the walls, use toggle bolts to fasten the units to the walls.*

6. *To fasten one unit to another, clamp the units together, align them, and then attach them with bolts and T-bolts.*

FACTORS TO CONSIDER

- Have plenty of support or help when installing cabinets.
- Be sure that you have found the studs and have fastened the cabinets to them.
- If cabinets are not well fastened, they will fall; if they are not level, drawers will stick and doors will not shut correctly.

LAYOUTS

Creating an accurate master layout is the key to the successful construction of cabinets. The layout contains all of the elements required in planning the location and design of the cabinets.

Basic Procedures in Developing Master Layouts

1. *Master layouts can be laid out on plywood or cardboard. Several layouts will be required. These layouts will be section views of the structure including clearances needed for drawers, pull-out boards, and special shelving.*

2. *Draw each member full size. Follow the overall dimensions provided in the architectural plans and details. Show clearances that may be required.*

3. *Use the master layout when laying out the exact sizes and locations of drawer parts and other detailed dimensions not included in the regular drawings.*

4. *Make any changes needed and proceed with construction.*

SHELVES

Shelves must be sturdy and suitable for holding a variety of items. If they are exposed, they should have a finish that complements the cabinets in which they are housed.

Basic Procedures for Installing Cabinet Shelves

1. *If the shelves are to be permanently set in the cabinets, cut dado joints in the side panels and glue the shelves in these joints.*

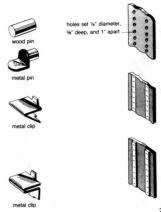

holes set ¼" diameter, ⅜" deep, and 1" apart

wood pin

metal pin

metal clip

metal clip

(Figure 4-13)   Adjustable shelf supports

2. *Adjustable shelves provide the user with greater flexibility in filling cabinet space. Several systems can be used to hold adjustable shelves in such a way that they are easy to move but sturdy while in use. Two of the systems require that holes be drilled in the support panels and either wooden pins or metal shelf pins inserted in the holes; the shelves rest on the wooden or metal pins. Other systems use metal strips that fit in groove cuts in the side panels or that are fastened on the surfaces of the side panels. (See Figure 4-13.)*

3. *Lay out the shelf support system so that the shelves will be level.*

4. *If the shelf support system chosen requires grooves to be cut in the side panels, this must be done before the cabinet is assembled.*

5. *The standard thickness for shelving is ¾ inch. This shelving requires support every 32 inches or less, depending on the load the shelf will be carrying.*

FACTORS TO CONSIDER

- If plywood is used for shelving, face it with solid stock that matches the cabinets.

- Do all necessary cutting and drilling before the cabinet is assembled for both better accuracy and ease of construction.

CEILINGS

Ceilings serve three primary purposes within a structure: structure support, sound control, and decoration.

CEILING MATERIALS

Ceilings are finished with a number of different materials based on function and personal preference. The materials most commonly used are raw materials such as drywall upon which a number of finishes or textures can be applied. Finished materials that can be installed with no further work needed can also be applied.

CEILING TILES

Ceiling tiles are used in both old and new construction. They can be installed over almost any surface that has been made even, smooth, and continuous, and they come in a variety of colors and textures.

Tiles are made from many different materials, but the most common are fiberboard, mineral, perforated metal, fiberglass, and plastic. The selection of the tile material should be based on appearance, light reflection, fire resistance, sound absorption, maintenance, cost, and ease of installation. The standard size for ceiling tile is 12 × 12 inches. However, tiles are available in larger sizes such as 24 × 24 inches and 16 × 32 inches. The larger tiles are used primarily in commercial settings.

Basic Procedures for Installing Ceiling Tiles

1. *Measure the two short walls and locate the midpoint of each one. Snap a chalk line to establish a line that is at right angles to the long centerline. All tiles are installed with their edges parallel to these lines.*

2. *When the ceiling measurements do not match the tile dimensions to allow full tile to be set throughout the ceiling, split the difference equally between the two border courses. For example, if there is a 10-inch difference, allow 5 inches on either side of the ceiling. This will give an even visual balance.*

(Figure 4-14)   Furring strips

3. *Furring strips should be nailed to the ceiling joists. The furring should be applied at right angles to the joists. The first furring strip should be nailed against the wall. The second strip should be placed so that it is centered over the edge of the border tile. Use the chalk line that was previously laid out. All other strips are then installed O.C. (on center) for the size of tile being installed. The furring strips must be level with each other; if they are not, use shims to make them level. (See Figure 4-14.)*

4. *Carpenters have differing preferences regarding where to start installing the tile. Some start from the center and work each side toward the walls. Others start with the border tiles and work from one side to the other. If the furring strips are even and uniform, the center-to-sides method is the faster one.*

5. *If you encounter electrical or plumbing lines that are only slightly below the joists, work around them by adding a second layer of furring strips. The second layer of furring should be installed at right angles to the first layer. When pipes or wires are encountered, the furring should be notched to fit around them. Where electrical or plumbing lines are encountered far below the ceiling, you should box around them with furring strips. Wood or metal strips can be used to finish the corners or edges.*

6. *Another method of installing tile is to place drywall sheets over the joists and then glue tiles to the drywall. This method is used when the texture of the tile is desired.*

7. *The tile is installed by adhesive or staples. The adhesive is applied in each corner of the tile, and the tile is pressed into position. When using staples, three staples are used along each edge of the tile while it is held in position. The length of the staples should be at least 9/16 inch for sufficient holding power.*

8. *A tile can be cut by scoring it with a knife, placing it over a sharp edge, and then applying downward pressure. This will cause the tile to break along the score line.*

9. *A third method of attaching ceiling tile is the metal track system. This system uses 4-foot metal tracks instead of wooden furring. The metal is nailed directly to the joists, and the tongue and groove tiles are slipped into place. A clip snapped into the track slides over the tile lip. No other fastener is required.*

DRYWALL

Drywall is the common name for gypsum wallboard. It is laminated material with a gypsum core and paper covering on each side. It usually comes in 4 × 8-foot sheets but is also available in 6, 7, 9, 10, 12, and 14 foot lengths. It is sold in the following thicknesses: ¼, 5/16, ⅜, ½, and ⅝ inch. Consult your local building codes for specifications on minimum thicknesses required for various applications. Thin drywall can be purchased for use in creating arched shapes, such as window arches and entry ways. Greenwall is special water resistant drywall that is used in moist areas such as bathrooms. Drywall is used on both walls and ceilings. Drywall is manufactured

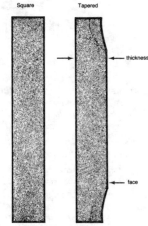

(Figure 4-15)   Drywall types

are tapered and square. The tapered edges create a shallow channel between the sheets into which joint compound can be easily placed and brought to level. This creates a smooth, even surface. (See Figure 4-15.)

Basic Procedures for Installing Drywall Ceilings

1. *The thickness of drywall used in quality construction is ⅝ inch. Use the longest drywall sheets possible to eliminate extra joints.*

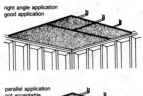

right angle application
good application

2. *When hanging drywall, always do the ceilings first, and then do the walls.*

3. *Place the drywall sheets at right angles to the ceiling joists. (See Figure 4-16.)*

4. *All measurements should be taken from where the sheet will be installed.*

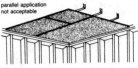

parallel application
not acceptable

(Figure 4-16)   Installing drywall

5. *Drywall can be cut across the width or length of the board by first making a scoring cut with a knife pulled along a straightedge. The scoring cut should cut through the paper and into the gypsum core. When the scoring cut is completed, the main part of the board should be supported and the excess stock snapped by applying downward pressure. When the gypsum core has been broken, complete the cutting by drawing the knife through the back paper.*

6. *Irregular shapes and curves can be cut with a coping saw, compass saw, or electric saber saw.*

7. *Drywall is secured to the joists with annular ring nails, commonly called sheetrock nails, or wallboard screws. Either of these fasteners provide a firm, tight fit to the wood joist. If metal stringers are being used, fasten the sheetrock with self-tapping screws that are made for metal. The sheetrock nails should be driven straight into the wood. The nailhead should be driven to rest in a slight dimple below the surface. Take care in making the last few hammer strokes so that you do not break the paper face of the board.*

8. *When applying drywall to the ceiling, make sure you have a platform*

on which to stand, or use stilts, so that you can reach the work easily.

9. *The drywall itself can be supported with T-supports. These T-supports are adjustable for different ceiling heights. If T-supports are not available on the job, they can be easily made out of 2 × 4s.*

10. *The next step is to apply the joint compound. If you are going to do the taping and joint work, refer to the section on drywall finishing in this book for specific step-by-step instructions.*

11. *When the joints have been completed, the ceiling is ready to be finished. If the ceiling is to be painted, follow regular painting procedures. A texture can be sprayed on the ceiling or applied with joint compound to create a random design by swirling a brush or broom into the wet compound. If the spray method is used, it's just a matter of knowing how to operate a spray gun. On most jobs, the drywallers come in and do this operation. If you have to do the spraying yourself, the special machines and materials that you will need for the texture spray are available from most rental firms. If you create the random-design texture, you will need to practice making swirls with a brush or broom. With a little experience, you will be able to create a ceiling surface that is decorative and functional.*

SUSPENDED CEILINGS

Suspended ceilings are commonly used in commercial settings or in homes where heating ducts or plumbing lines interfere with the application of a finished surface. The height of the original ceiling must be high enough to allow for a dropped ceiling of standard height.

Basic Procedures for Installing Suspended Ceilings

1. *Determine the exact height of the desired ceiling.*

2. *Attach the metal molding around the perimeter of the room. Use a chalk line to assist in the layout process. Be sure that the molding is securely fastened into the studs. If the walls are concrete, use a power gun to secure the molding.*

3. *Lay out the room dimensions and position the main runners for assembly. Attach the runners to the joists with wires run through the screw eyes that are located 4 feet O.C. Make sure that the screw eyes and wires are fastened tightly to each other and the joists to prevent the ceiling from sagging.*

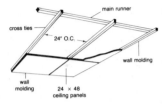

(Figure 4-17)   Suspended ceiling

4. *Using the same calculation used for regular ceiling tiles, determine the size of the side panels if the room dimensions dictate.*

5. *Once the main runners are in position, check for level and adjust any of the holding wires until level is achieved.*

6. *Install the cross ties. This is done by inserting the end tab of the cross tie into the runner slot. The suspension framework is now complete.*

7. *Install the panels. They can be installed by tilting each one upward and turning it slightly on edge so it will "thread" through the opening. When the entire panel is above the framework, turn it flat and lower it onto the grid flanges. Installation is now complete.*

8. *Some suspended ceilings have hidden interlocking panels. The concealed suspension system is installed in the same manner as a regular suspended ceiling. (See Figure 4-17.)*

STRUCTURE SUPPORT

Structure support is the most important function of ceilings. Without the proper support, the structure will be unsound and thus unusable. Structure support is achieved by framing in the ceiling. Ceiling framing has three main purposes:

1. Tying together opposite walls and roof rafters to resist outward pressure on walls imposed by pitched roofs.

2. Supporting finished ceilings.

3. Supporting second story or attic storage areas.

Ceiling framing is similar to floor framing. The main differences between the two are that ceiling joists are lighter than floor joists and header joists are not included around the outside of the ceiling joists. The ceiling joists do not carry the weight that the floor joists do, and thus the load capacity does not have to be as great. On some multistory structures, the bottoms of the floor joists carry the ceiling cover. When trusses are used for roof framing, a ceiling frame is not required. The bottom chords of the trusses carry the ceiling surface.

The joists are the main ceiling framing members. The joist size is determined by their spacing and length of span. Local building codes must be checked for minimum specification requirements. As with all other house framing, 2-inch stock is used.

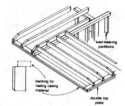

(Figure 4-18)   Joists

Ceiling joists are normally put on 16-inch centers in order to permit the use of a wide range of ceiling materials. (The architectural plans will give the exact O.C. requirements.) The joists normally run across the narrow dimensions of a structure. In some cases, some joists can run in one direction and others at right angles. The change in angle occurs at a junction over a load-bearing partition wall. (See Figure 4-18.)

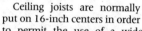

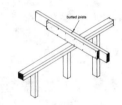

(Figure 4-19a)   Framing a ceiling

Basic Procedures for Framing Ceilings

1. *Ceiling joists can be joined over load-bearing partitions by using a lapped joint, where the joists are nailed to each after they have been overlapped, or by using a butt joint with a scab board nailed along one side. (See Figure 4-19(a) + (b).)*

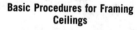

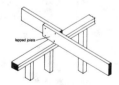

(Figure 4-19b)   Framing a ceiling

2. *Lay out ceiling joists by marking both the outside plates and the interior wall partitions. The rafters frame opposite one another, like ribs, down the center line of the house (the main beam). The ceiling joists run from the outer wall of the house to the load-bearing partition inside the house. Because the joists begin alongside the rafters, there will be a gap between the joists where they overlap*

atop the load-bearing partition. To fill this gap, place a filler block between the two joists. The filler block is the same width as the rafter and thus fills the space perfectly. (See Figure 4-20.)

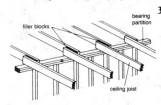

(Figure 4-20)   Filler blocks

3. *Ceiling joists that are cut for outside walls must be trimmed to match the slope of the rafter, since the width of the ceiling joist at the outside wall will be wider than the rafter. This extra width, if not removed, will interfere later with the installation of sheathing. The layout for this trim job can be done in one of two ways: (1) use a framing square to establish the rise and run, mark the excess, and saw it off or (2) when the amount of excess stock is small, it can be removed with a hatchet or saw. (See Figure 4-21.)*

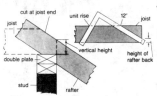

(Figure 4-21)   Trimming a joist

4. *Ceiling joists are installed before the rafters.*

5. *Ceiling joists are toenailed to the top plates of the exterior walls with two 10d nails on each side.*

6. *At the center lap, the joists are nailed to the filler block and then toenailed to the plate of the load-bearing wall.*

7. *The joists must be well nailed since they are cross-tying the structure together.*

8. *Ceiling joists and nonload-bearing partitions that run parallel to each other are harder to anchor together because of the lack of nailing places. With minor framing, this problem can be overcome by nailing a 2 × 4-inch*

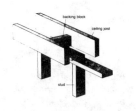

(Figure 4-22)   Backing block

backing block between the ceiling joists running across the double plate and then toenailing the block into the double plate. (*See Figure 4-22.*)

9. When running ceiling joists parallel to the walls, a nailing strip or drywall clip must be installed so there will be something there to which the ceiling material can be fastened. The chief requirement for this nailing strip is that it provide adequate support. (*See Figure 4-23(a) + (b).*)

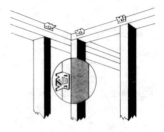

10. Nearly every structure will have an attic access opening. The attic access should be incorporated into the ceiling framing in order to ensure sufficient structural strength. Fire regulations and local building codes usually list minimum size requirements for an attic access. The building plans generally indicate specific size, description, and location of the opening. The opening is framed similarly to a floor opening. If the size of the opening is small (2 to 3 feet square), doubling of the joists and headers is not required. (*See Figure 4-24.*)

(Figure 4-23b)   Drywall clip

(Figure 4-23b)   Nailing strip

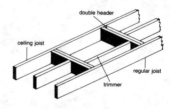

(Figure 4-24)   Attic opening

11. On long ceiling spans, it is good practice to use strongbacks for additional support, to maintain the correct spacing between the joists, and to even up the bottom edges of the joists so that the ceiling will not be wavy after the drywall is applied. A strongback is an L-shaped support that is attached across the tops of

the joists in two steps. The first step is to take a 2 × 4 and mark off the proper spacing (16 or 24 inches O.C.). Place the 2 × 4 across the ceiling joists and fasten it with two 16d nails at each joist. Apply

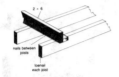

(Figure 4-25)   Long ceiling span

pressure against the joists as needed to bring them into proper position. The second step is to select a straight 2 × 6 or 2 × 8 for the vertical part of the L. Place this member against either side of the 2 × 4 that has already been attached to the joists. Attach the member to the 2 × 4 with 16d nails. Hammer the nails in-between, not over, the joists. Work across the full length of the strongback, aligning and nailing. Placing your foot on either the 2 × 4 or the edge of the vertical member will help align each joist. Once each joist is aligned, toenail the vertical member to it. (See Figure 4-25.)

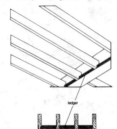

(Figure 4-26)   Joist ledger

SPECIAL CEILING FRAMING

Plank and beam ceilings are commonly used in construction,

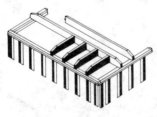

(Figure 4-27)   Stub joists

especially in A-frame and recreational or seasonal homes. This form of ceiling eliminates the need for ceiling joists. The roof planks serve as the finished ceiling. The planks are selected for appearance, and once in place the only additional work needed is to finish the wood. Beams are used to support the ceiling and to distribute the

FACTORS TO CONSIDER

- When framing large rooms, the midpoints of the joists may need to be supported by a beam. This beam can be located below the joists or installed flush with the joists. In the latter installation, the joists may be carried on a ledger. (See Figure 4-26.)

- The beams can also be secured by using joist hangers. These metal hangers come in various widths and heights. Selection is based on the number of 2-inch pieces used for the beam and the width of stock (6 inches, 8 inches, etc.).

- Since ceiling joists serve as supports and anchor together wall partitions and outside walls, care must be taken to ensure that the joists are securely fastened.

- On low-pitched or hip roofs, it is necessary to stub ceiling joists along the end wall. These short pieces or stubs will give clearance for the rafters and get the ceiling joists going in the direction of the hip rafters, giving support and a nailing surface to the rafters. (See Figure 4-27.)

- Always keep in mind when framing a ceiling that you must provide a place for attaching ceiling materials. Nailing strips or drywall clips must be in place. This will save time and frustration later on.

weight to walls and ultimately to the foundation. Certain beams are "spaced" beams or hollow beams in which the electrical and plumbing distribution systems are concealed. These spaced beams are created by using several pieces of 2-inch stock separated by short blocking. (See Figure 4-28.) The specifications for these beams are usually given in the architectural plan.

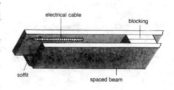

(Figure 4-28) Special ceiling

DOORS

EXTERIOR DOORS

Exterior doors create barriers between the interior of a structure and the outside elements. These doors are composed of a number of different materials and operate in a number of different ways. All types, however, must provide protection from noise, weather, and unauthorized entrance. They must be attractive, easy to operate and maintain, and made of materials that will last while exposed to harsh elements. The primary materials used in the construction of doors are metal, wood, glass, and combinations of two or all three of these.

DOOR FRAMES

Door frames hold the door in position during its operation. These frames have three parts: a left side, a right side, and a top. These parts are known as jambs and are designated by their location, such as left jamb or top jamb. Exterior doors usually have a horizontal piece of hardwood at the bottom of the opening called a sill. On top of this sill is placed a shaped piece of hardwood or a plastic and metal strip called a threshold. The threshold covers the joint between the sill and the flooring material and provides a seal to keep out moisture and weather, while setting the door up high enough to clear the interior floor covering.

Door frames are made of wood, solid steel sections, hollow aluminum, or hollow cold-rolled steel. The installation procedure depends on the type of frame that is being used.

Basic Procedures for Installing Door Frames

1. *Metal door frames come with jambs and heads ready to be set into place. Adjustments can be made for level and squareness with screws that are in the head of the frame. These frames may come in a kit with specific directions for installation; if so, follow the directions carefully.*

2. *Before you actually start positioning the door frame, check that a trimmer is set on either side of the door opening and that a double*

header is located across the top. The trimmers and headers must be in place to help span the door opening. Without them, the wall could sag and bring pressure on the door frame, adversely affecting the operation of the door.

3. *If the door frame is to be placed in an opening created in a masonry wall, metal anchors should be built into the wall. If these anchors were not put in during the construction of the wall, you will have to anchor the door frame to the wall with anchors or nails driven by a power gun. On some masonry walls, the opening may be framed with 2 × material that is of the same width as the masonry wall. Commonly, this will be 2 × 6 lumber. The door frame then can be fastened to the 2× material, providing a strong nailing surface.*

4. *Wood door frames come with the side jambs and head or top jamb. The jambs are sanded and ready for assembly. They should be treated with as much care as finished woodwork.*

5. *Before the door frame is set into place, check the opening for final measurements on the sides and header to see whether the frame has to be trimmed to fit. When standard measurements are used, the jambs should not have to be cut to fit, but measuring will same time in case there has been a change during construction of the door opening.*

6. *Nail the jambs together using 8d casing or box nails.*

7. *Set the frame into the opening. Let the side jambs rest on spacer blocks that are the thickness of the finished floor.*

8. *Level the head jamb by trimming the side jambs as needed.*

9. *To hold the frame firm while the side jambs are being squared, place a 1 × 6 spreader between the side jambs at the floor level. The spreader's length should equal the distance between the side jambs measured at the top.*

10. *Center the frame in the opening. Secure it with double-shingle wedges at the top and bottom on each side. Plumb the jambs with a straightedge and level or a long carpenter's level. Make adjustments in the double-shingle blocking until each side is in correct position. Then fasten the top and bottom of each side jamb with an 8d casing nail.*

11. *Continue blocking the frame up and down the jambs. On the hinge* *jamb, position one block 11 inches up from the bottom and another block 7 inches down from the top. Set a third block halfway between these two. Also, set a block at the latch level on the latch jamb. Make a final check that the frame is square and level before nailing. Use 8d casing nails to complete the nailing, and be sure to nail through the spacing blocks. If you nail where there are no blocks, you can bow or pull the jamb out of square because there is no support behind it.*

DOOR HARDWARE

If the unit is prehung, normally the openings for the hardware, such as door locks and deadbolts, have already been cut. If this is the case, then it is a matter of assembling the hardware through the openings and making sure they work properly.

If the door is not prehung and you have to install all of the hardware, you first must select the correct hardware. In some cases, you will need to specify the "hand of the door" or the direction of the door swing.

Basic Procedures for Installing Door Hardware

1. *When the hardware has been selected (door lock, deadbolt), open the package and read the directions carefully. Many packages include a template that can be used for drilling the lock openings. Open the door to a convenient position and block with wedges so it will not move. Measure up from the floor a distance of 38 inches and mark a horizontal line on the edge. This will be the center of the lock. Position the furnished template on the face and edge of the door. Lay out the center of the holes. The height for some locks is 36 inches. Check this height before drilling the lock holes.*

2. *Proceed to drill the proper size holes in the correct location.*

3. *If you are installing a number of door locks, get a boring jig. Boring jigs assure accuracy, save time, and eliminate the use of the lock templates.*

4. *The shallow mortise on the edge of the door can be laid out and cut with wood chisels.*

5. *For the faceplate mortise, wood chisels can be used. If a number of*

locks are being installed, a faceplate mortise marker or marking chisel will make the job go faster and more accurately.

6. *Assemble the hardware and check to see that all the pieces operate smoothly and correctly.*

EXTERIOR DOOR THRESHOLDS AND DOOR BOTTOMS

Exterior doors require thresholds to seal the space between the bottom of the door and the door sill. Most thresholds are made of metal and rubber or vinyl sealing strips. Thresholds come in many different designs and colors. The vinyl strips fit into special channels and are replaceable when they wear out.

Basic Procedures for Installing Thresholds

1. *Prehung exterior doors come with thresholds already installed. The only adjustment that is necessary is to set the height and level of the threshold. This is done by moving the threshold up or down with screws located beneath the vinyl strip on either end of threshold. The manufacturer's instructions will help in this effort.*

2. *For doors that are not prehung, the threshold must be cut to fit the opening left after the door frame has been set into place. The threshold should be cut so that it fits snugly inside the door frame.*

3. *Fasten the threshold in place with the screws that are furnished with the threshold.*

4. *Swing the door shut and check the seal that is created between the vinyl strip and the door bottom. Adjust the threshold up or down until the proper seal has been achieved without having to force the door shut. If the seal is too tight, the vinyl will tear out in a very short time.*

5. *Thresholds are commonly thought of as metal and vinyl strips set to meet the bottom of the exterior door. When hanging doors another form of threshold must be installed and this is called a saddle. The saddle is the part of the threshold that slips in between the door frame and creates a lip on the exterior side of the door. The saddle comes as a part of the door frame kit and is installed when the frame is put in place. It is on this saddle that the metal threshold is set.*

PREHUNG EXTERIOR DOORS

Prehung exterior doors come set in their frames ready to be set into place. They have support strips nailed from jamb to jamb to hold the door square during shipping. These strips should be removed just before setting the door in place to reduce the chance of knocking the door out of square. If the door frame is metal, the support will be across the bottom, since metal is more rigid and less likely to move.

The opening for the door should be exact so that the door frame will not have to be trimmed. Prehung doors come with the threshold installed, so, if trimming is necessary, try to make the changes on the opening rather than on the frame, since the frame will have to be disassembled, cut, and reassembled.

The exterior side of the frame comes with the casing already in place, so leveling and squaring will be done from the interior.

Basic Procedures for Installing Prehung Exterior Doors

1. *Generally, the door assembly will be square. You will only need to bring the assembly to level by using shims under the threshold.*

2. *Place shims along the vertical jambs to support the frame. These shims should be located at the four corners of the frame, the hinges, and the latch area.*

3. *Nail the frame in place using 8d casing nails.*

4. *Prehung doors come with predrilled holes for the hardware; all that is needed is to install the latches, locks, or deadbolts.*

TYPES OF AND MATERIALS FOR EXTERIOR DOORS

Exterior doors come in a variety of materials and styles. (See Figure 4-29.) The materials used to construct the door and the door style depend not only on the degree of protection or privacy desired, but also on such factors as architectural compatibility, esthetics, fire resistance, and cost.

Materials used to manufacture exterior doors include wood, metal, plastics, glass, and various combinations of these. Also, manufacturers are utilizing technology to produce doors with such features as low-emissivity glass for improved home weatherization.

The standard widths for exterior doors are 2 feet 6 inches; 2 feet

8 inches; 3 feet; and 3 feet 6 inches. Standard heights are 6 feet 8 inches and 7 feet.

Swinging doors are the most commonly used type of door and are classified as either right hand or left hand, depending on which side they are hinged. The left and right sides of the door are determined by standing outside of the door facing it. (See Figure 4-30.)

Metal and Glass Doors. Metal doors for most construction are classified as hollow core. They are usually 1¾ inch thick and made of cold-rolled steel with a factory finish. They may be either flush or panel in design. They normally come prehung or, if not, the hinge and hardware openings are stamped in and ready for installation. The core of the metal door is usually filled with foam or some other

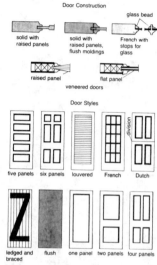

(Figure 4-29) Exterior doors

form of insulation to conserve energy and deaden sound. Depending on where these doors are located, they may have glass inserts or plastic overlays. They can be painted to match any color scheme being used, and decorative moldings and trim can be applied to match the door with its surroundings.

Basic Procedures for Installing Metal and Glass Exterior Doors

1. *Metal doors generally come prehung and thus are set into the door opening with no installation of hinges needed. Squaring of the door frame is accomplished as discussed in the section on door frames. If for some reason the door does not operate smoothly after installation, check to see if the frame is sprung or the door is twisted. If either of these conditions is present and the door frame is metal, you can try to straighten the frame through shimming between the frame and the casing edge, if you can get between them. Generally, the only way to*

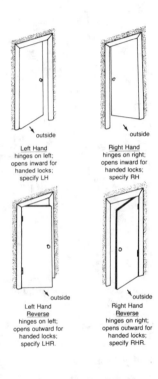

Left Hand
hinges on left;
opens inward for
handed locks;
specify LH

Right Hand
hinges on right;
opens inward for
handed locks;
specify RH

Left Hand
Reverse
hinges on left;
opens outward for
handed locks;
specify LHR.

Right Hand
Reverse
hinges on right;
opens outward for
handed locks;
specify RHR.

(Figure 4-30) The hand of a door

correct this problem is to return the unit and order another, since you can't recut the hinge gains or make other modifications in the metal frame.

2. *Glass doors require a different form of frame and hinging system. There are many forms of glass doors, and the best way to install them is to read and follow carefully the manufacturer's directions. Glass doors must be in exact alignment to operate properly. There is minimum opportunity for adjustment once the door is set in place, so all preceding measurements and construction efforts must meet the specifications.*

Wood Doors. Wood doors are constructed primarily of solid pine. This gives the door stability and beauty. The doors may come prehung or as a separate unit requiring hanging. Wood doors come in a variety of styles ranging from flush to paneled. They may also be customized by the application of decorative moldings and trim.

Basic Procedures for Hanging Wooden Exterior Doors

1. *Make sure the door is the correct one for the opening.*
2. *Determine the direction that it will swing.*

3. *Lay out and mark the door jamb that will receive the hinges.*

4. *Mark the edge of the door on which the hinges will be mounted.*

5. *Size the door to see whether it will need to be trimmed. There should be only a small amount of trim work needed if the opening has been made to standard size. Clearance should be $3/32$ inch on the lock side and $1/16$ inch on the latch side. The top should have $1/16$-inch clearance; the bottom, $5/8$ inch. If weatherstripping is used around the door, the clearance should be increased by $1/8$ inch. The threshold height will determine the exact clearance needed on the bottom. The threshold can be installed before or after the door is hung. If possible install the threshold first in order to determine the exact clearance needed on the bottom.*

6. *The most common method of cutting gains for the hinges is to use a door-and-jamb template with an electric router. This method saves time and effort. The template is positioned on the door and the cutout sections set where the hinges are to be installed. The gains are then cut out. The template is then attached to the door jamb where the matching cuts are made. Adjustments should be made where needed based on the door thickness and height. The template is so constructed that it is almost impossible to cut the gains on the wrong side of the door or jamb, but a second check never hurts.*

7. *Most hinge templates are designed to cut gains for round hinges. If installing square hinges, square the corners with a chisel.*

8. *If using a chisel to cut the gains, follow this procedure. Place the hinge on the door and mark around the outline. Score the hinge outline with the chisel inside the pencil mark. Hold the chisel vertically and drive it to the depth of the hinge width. Keep the beveled face of the chisel toward the opening. Make shallow feather cuts to clean the gain to the depth line. Hold the chisel's beveled edge downward. Make small cuts, since this will create a cleaner gain. Next, hold the chisel along the depth line and tap lightly across the grain of the gain to shave off the feather cuts. Hold the chisel with the beveled side of the blade facing up. Clean the gain of any feathers or uneven wood slivers.*

9. *Place the hinge in the gain so that the head of the removable pin will be up when the door is hung. Drive the first screw in slightly toward the back edge to draw the leaf of the hinge tightly into the grain. Follow this same procedure to attach the free leaf of the hinge*

to the jamb. Many carpenters prefer to set only two screws, check the fit and then proceed with the rest of the screws.

10. *Hang the door and check for fit. The door can be trimmed if needed, or the depth of the hinges can be adjusted by cutting the gains deeper or shimming them out, depending on the adjustment needed. If shims are needed, the best material to use is cardboard, since it is easy to cut and stands up well to pressure without compressing too much.*

FACTORS TO CONSIDER

- When cutting hinge gains, the depth of the router bit needs to be checked very carefully. The depth of the gain plus the thickness of the template can sometimes throw off the measurements.

- If using a chisel to cut the hinge gains, make sure the tool is sharp and that you have control of it at all times because it is very easy to slip and split the edge of the door or jamb.

INTERIOR DOORS

Interior doors create barriers between the various parts of the interior of a structure. These doors are used for privacy, noise reduction, and decoration. They are made from a wide variety of materials, although the most commonly used are wood and plastic. Interior doors, like exterior doors, come in many forms to fit the decor and meet the needs of the structure in which they are located.

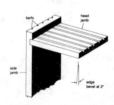

(Figure 4-31)   Door frame

DOOR FRAMES AND CASINGS

Interior door frames serve the same purpose as exterior door frames, that is, to hold the door in position during its operation. These frames have three parts: a left side and right side, called jambs, and a top, called a head jamb. Interior frames are simpler than exterior frames. The jambs are not rabbeted, and no sill is included. (See Figure 4-31.)

Standard jambs for 2 × 4 stud partitions are made from 1-inch material. For drywall partitions the jambs are 4½ inches wide. The back is usually kerfed to reduce the tendency toward cupping. The front edge of the jamb is beveled slightly so that the casing will fit tightly against it with no visible crack.

The side jambs are dadoed to receive the head jamb. The side jambs for residential doorways are made 6 feet 9 inches long (measured to the head jamb). This provides clearance at the bottom of the door for flooring materials.

Some interior doorjambs are designed to be adjustable to meet different wall thicknesses. One style uses a rabbet joint and a concealing door stop; another is made of two pieces with the door stop located over the crack that is left after adjustment. (See Figure 4-32.)

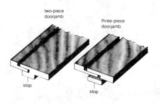

Door casing is applied on each side of the door frame to cover the space between the

(Figure 4-32)   Doorjamb

frame and the wall. It also supports the jambs so that they can carry the door, and secures the frame to the wall.

Basic Procedures for Installing Interior Door Frames

1. *Interior door frames are normally made from wood when used in residences. For use in other structures, the frames may be wood or metal. Metal frames are set into place and adjusted using set screws as directed by the manufacturer's instructions.*

2. *Check that the opening is the right size and correctly completed. When standard measurements are used, it isn't necessary to cut the jambs to fit—they will set right in. It is always a good idea to check the measurements before you start to work in case there has been a change during the construction process.*

3. *After the jamb parts have been laid out and made ready for installation, nail the frame together using 8d casing nails.*

4. *Set the frame into the opening. Let the side jambs rest on spacer blocks that are the thickness of the finish floor.*

5. *Level the head jamb by trimming the side jambs as needed.*

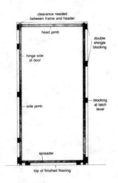

(Figure 4-33) Installing a door

6. *To hold the frame firm while the side jambs are being squared, place a 1 × 6 spreader between the side jambs at the floor level. The spreader's length should equal the distance between the side jambs measured at the top.*

7. *Center the frame in the opening. Secure it with double-shingle wedges at the top and bottom on each side. Plumb the jambs with a straightedge and level or a long carpenter's level. Make adjustments in the double-shingle blocking until each side is in correct position. Then fasten the top and bottom of each side jamb with an 8d casing nail. (See Figure 4-33.)*

8. *Continue blocking the frame up and down the jambs. On the hinge jamb, position one block 11 inches up from the bottom and another block 7 inches down from the top. Set a third block halfway between these two. Also, set a block at the latch level on the latch jamb side. Make a final check that the frame is square and level before nailing. Use 8d casing nails to complete the nailing, and be sure to nail through the spacing blocks. If you nail where there are no blocks, you can bow or pull the jamb out of square.*

9. *To cover all the nail holes, measure in from the door side of the jamb a distance equal to the door thickness plus ⅞ inch. Make a light pencil line around the jamb, and nail on this line. The door stop will cover all the nails when it is installed.*

Basic Procedures for Installing Door Casings

1. *Lay out the needed pieces.*

2. *On the edge of the jamb, draw a light pencil line ¼ inch back from the face.*

3. *Be sure that the bottom end of the side casing is square so that it will rest tightly on the finished floor.*

4. *Hold the side pieces tightly in place, and mark the miter joint at the top. Use a miter box and backsaw to cut the joint.*

5. *Nail the side casing temporarily with finishing nails. Mark, cut, and fit the head casing. If the miters do not fit properly, trim them with a block plane.*

6. *Complete the nailing process. Use 4d or 6d nails along the jamb edge and 8d nails on the outer edge into the studs. Each pair of nails should be about 16 inches O.C.*

7. *Set the nails. If hardwood casing is used, it is advisable to drill holes first.*

PREHUNG INTERIOR DOORS

Prehung interior doors come set in their frames ready to be set into place. They have support strips nailed from jamb to jamb to hold the door assembly square during shipping. These strips should be removed just before setting the door in place to reduce the chance of knocking the door assembly out of square.

The opening for the door should be exact so that the door frame will not have to be trimmed.

Interior doors normally swing into the room that they serve. The prehung door assembly should be opened on the room side of the rough opening.

Basic Procedures for Installing Prehung Doors

1. *Study the manufacturer's directions to find out whether there are any special procedures that should be followed.*

2. *Remove the casing that is not attached to the door frame and set it aside.*

3. *Set the door frame (the part with the door) into the opening. The side jambs should rest on the finished floor or spacer blocks.*

4. *Plumb and square the door frame. Nail the casing to the wall structure.*

5. *On the other side of the wall, install shims between the side jambs and the rough opening. Nail through the jambs and shims. Make sure you hit the shims each time you nail, or you may draw the frame out of square.*

6. *Install the remaining half of the frame. Make sure the tongue edge is inserted into the grooved section already in place. Nail the casing to the wall structure. Some doors do not include the casing; with these, only the door frame would be nailed.*

7. *Finish nailing the frame in place. Remove the spacer blocks, and check the door operation. Make adjustments as needed.*

8. *Install door hardware.*

FACTORS TO CONSIDER

- Check and recheck for square. The slightest variance in the frame can mean that the door will stick or bind.

- Make sure enough shims are placed between the frame and the wall opening. If adequate support is not in place, the frame will move or warp, causing the door to stick or not close properly.

TYPES OF AND MATERIALS FOR INTERIOR DOORS

Interior doors are normally made of wood, plastic, metal, or glass. The glass doors are specialty doors used primarily in business settings, and the metal doors are used for heavy-use areas and fire protection. The standard thickness for interior doors is 1⅜ inches. The widths for interior doors are 2 feet 6 inches; 2 feet 8 inches; and 3 feet. Doors used to enclose special areas such as closets may be 2 feet wide. Doors can be made to your specifications, based on structure and personal preference.

Both interior and exterior doors are graded and manufactured according to standards developed by the National Woodwork Manufacturing Association (NWMA) and the Fir and Hemlock Door Association (FHDA). These grades and standards were developed to establish nationally recognized dimensions, designs, and quality specifications for materials and work. The designations of these grades and standards can be obtained by contacting the above-mentioned associations.

There are two common forms of interior doors used: panel and flush. Panel or framed doors have vertical and horizontal members that frame rectangular areas in which trim, glass, or louvers are located. Panel doors are solid core, since they are constructed from solid materials. The sides of a panel door are called stiles. The stiles run the full length of the door. The horizontal members of the door are called rails. Normally, three rails are used in a door. The top and bottom rails are used to support the door, and the middle rail, called the lock rail, supports the door but also provides a surface on which to attach the lock. Some panel doors have mullions. These are relatively heavy vertical members that subdivide the areas between the stiles and rails. The areas between the rails, mullions, and stiles are panels. These panels may be carved wood, plastic, glass, or ventilation louvers.

Flush doors very widely in their construction, but all types include face panels, which form the faces; a solid or hollow core, to which the face panels are attached and which gives them support; and edge strips, which surround the rim of the core to provide finished edges to match the material in the face panels. Hollow-cores may include vertical or horizontal ribs, rectangular grids, or small cells. Some cores include stiles and rails. Cores usually include solid blocking for attachment of locks and other hardware. Openings may be made in flush doors for the insertion of glass, for light or vision, or louvers, for ventilation.

Wood doors are constructed primarily of pine. In cheaper doors, the face of the door may be covered with plastic that has been colored and imprinted to simulate a certain kind of wood. Also the interior of hollow-core doors may have plastic strips. When working with cheaper doors, check that the manufacturer has provided wood inserts to which door hardware can be attached.

Basic Procedures for Hanging Wooden Interior Doors

1. *Make sure the door is the correct one for the opening. It is easy to choose the wrong door when many different ones are on the job site.*

2. *Determine the direction in which the door will swing and the side on which the hinges will be located. If the hinges are on the left*

side, the door is said to be a "left-hand swing." If the door swings inward it is called a "left-hand, single." If it swings outward it is a "left-hand, reverse single." A door that swings both ways through an opening is called a "double acting door."

3. Lay out and mark the door jamb that will receive the hinges.

4. Mark the edge of the door on which the hinges will be mounted.

5. Size the door to see whether it will have to be trimmed. There should be only a small amount of trim work needed if the opening has been made to standard size. Clearance on the lock side should be 3/32 inch and 1/16 inch on the latch side. The top should have 1/16 inch and the bottom 5/8 inch. If an extra thick floor covering is going to be used, the clearance should be increased accordingly.

6. The next step is to cut the gains. Number 6, on page 127, under the heading of "Basic Procedures for Hanging Wooden Exterior Doors" outlines this procedure.

7. The next step is to square the edges of the gain if needed. Number 7, on page 127, under the heading of "Basic Procedures for Hanging Wooden Exterior Doors" outlines this procedure.

8. If a chisel is being used to cut the gains, refer to the exterior door section that explains this operation.

9. See page 127, item 9, under the heading "Basic Procedures for Hanging Wooden Exterior Doors", for step-by-step instructions on placing the hinge in a gain.

10. Hanging the door and checking for fit is covered in item 10, page 128, under the heading "Basic Procedures for Hanging a Wooden Exterior Door".

DOOR HARDWARE

(See the section under Exterior Doors for directions on how to install door hardware)

DOORSTOPS

Doorstops are normally the last trim piece to be installed. Many carpenters like to cut and tack the stops in place before installing the lock. The stops help to guide the final location and installation of the locks.

Basic Procedures for Installing Doorstops

1. *With the door closed, set the stop on the hinge jamb with a clearance of 1/16 inch.*

2. *The stop on the lock side is set against the door except in the area around the lock. Allow a slight clearance for humidity changes and decorating.*

3. *Set the stop on the head jamb so it aligns with the stops on the side jambs.*

4. *The stops should be cut with miter joints and attached with 4d nails spaced 16 inches O.C.*

FLOORS

Floors are designed to support the weight of objects and to separate one level from another. The floor system and the structural frame are interrelated in that one element influences the other in the selection of the type of floor framing, composition, and materials. Local building codes must be met, and these codes vary tremendously throughout the country. Also, there are several alternatives for floor construction. These alternatives must be investigated so that the proper type of construction method is used to meet the needs of the particular project.

There are a number of factors that must be considered when laying out floor framing. These factors generally are considered by the architect who develops the plans for the building. But the same factors also must be considered by the carpenter when plans are modified or when a remodeling job is undertaken. Floor load is the first factor to consider. The floor must support the load that will be placed on it. This load factor is considered when determining the thickness of the concrete slab or the spacing and number of floor joists.

Sound transmission can be an important factor in certain settings. This concern can most often be solved with floor coverings or insulation, but there are times when floors must be constructed with space for dead air or for the insertion of sound-deadening materials.

The ceilings that are below the floors may influence the choice of flooring or floor framing materials. Wood joist construction or flat-slab concrete floors provide a smooth surface to which a ceiling can be readily attached. Floors with beams or girders may require the use of suspended ceilings if flat ceilings are desired. This decision regarding the materials that will be used for ceilings and floors must be made early in the construction process.

Whether the materials should be fire resistant also must be considered if the building is to be used for certain purposes that involve the risk of fire. If this is not a relevant consideration for the project, then ordinary wood-joist construction with wood subfloor may be used.

FLOOR COVERINGS

The selection of floor covering is one of the most important and yet difficult tasks of the construction process. The occupant of the structure will generally select colors and leave the selection of the composition of the flooring surface to the contractor. Such factors as appearance, wear life, upkeep, and cost are the basis for selection of most floor coverings. Floor-covering dealers can provide up-to-date information on developments in the flooring industry.

RESILIENT FLOORING

Resilient flooring includes vinyl, cork, rubber, and asphalt materials. The flooring comes in tile or sheet form and, with the exception of cork tile, can be installed anywhere in the house, including the basement.

Asphalt tile is the cheapest of the tiles. However, asphalt tile is seldom used today except on low-budget jobs. Asphalt does not wear well over the long term. The color of asphalt fades, and asphalt requires extensive maintenance and has limited grease resistance.

Vinyl tile is the most commonly used tile for a number of reasons. It is easy to install and maintain (it doesn't require waxing), and the color intensity lasts.

Cork tile is a natural material that has a pleasant warm appearance. It has a soft surface for walking and standing, and it is very good for sound control. Cork tile is easy to maintain if a surface coat of vinyl is applied to it.

Basic Procedures for Installing Resilient Tile

1. *Tiles are 9 or 12 inches square, and they are ⅛ or 1/16 inch thick. To estimate the number of tiles that will be needed for the job, first measure the length and width of the room in feet. For 12-inch tiles, simply multiply the length of the room by its width to determine the number of tiles required for the room. For 9-inch tiles, calculate the area of the room in inches and divide by 9. Divide irregularly shaped rooms into two or more rectangles, and calculate the number of tiles required as outlined above. When using two colors of tile, divide the total number of tiles required in half and buy equal quantities of each color. For complex patterns, lay out the tile required on graph paper and shade in the tile locations to determine the exact number of each color required. Buy extra tiles to allow for breakage and waste.*

2. *Prepare the subfloor so that it is smooth, dry, and clear of dust, dirt, grease, and wax.*

3. *Square off the room with chalk lines. To lay out and mark the room square, start by locating the center of the end walls of the room. Do not worry about breaks or irregularities in the contour. Mark the centerline by snapping a chalk line between the two points. Lay out another centerline at right angles to the main centerline. This can be done with a framing square. You can mark across the room with the framing square or lay out and snap a chalk line.*

4. *Make a trial layout of the tile in one section or quadrant of the centerlines. When you are near the wall, measure the distance between the wall and the last tile. If the distance is less than 2 inches or more than 8 inches, move the centerline closer to the wall by 4½ inches for 9-inch tile or 6 inches for 12-inch tile. This adjustment will eliminate the need to install border tiles that are too narrow. The border tiles must be exactly the same width on both sides of the room to provide a uniform appearance. Since the centerline is adjusted to exactly one half of the tile width, this adjustment ensures that the border tiles will be the same width.*

5. *With a spreader (brush or trowel), spread the adhesive over one of the quadrants. Spread the adhesive up to and even with the chalk line but do not cover it. Make sure the adhesive is spread evenly, without thin or thick spots. Let it set until it feels sticky but does not stick to you.*

6. *Start laying the tile at the center of the room. Make sure the edges of the tile align with the chalk line. Butt each tile squarely to the adjoining tile, with the corners in line. Lay each tile in place; do not slide it in place, since this can cause the adhesive to work up between the joints. Tile adhesive can be worked for several hours, so don't hurry. Make sure the tiles are where you want them before you set them down into the adhesive.*

Layout for Border Tile

Layout for Corner Tile

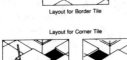

first marking   second marking

(Figure 4-34)   Tiling a floor

7. *Complete the main area. Border tiles are set as a separate operation.*

8. *To lay out border tile, place the tile that will be cut to fit over the last tile in the outside row. Then take another tile, place it in position against the wall, and mark a sharp pencil line on the first tile. (See Figure 4-34.)*

9. *Cutouts for corners are laid out in much the same way as border tile. Position a tile on top of the tile that is closest to the left-hand side of the corner on the last full row. Place another tile on top of this tile and slide it against the wall. Mark the first tile. This will give the width measurement. Next, reset the two tiles around the corner and mark the first tile. The two marks will intersect, showing the tile area that must be removed in order for the tile to fit into the corner. (See Figure 4-34.)*

10. *Cut the tile along the marked lines, using heavy-duty household shears or tin snips. Some types of tile require special cutters, or they may be scribed and broken. Asphalt tile can be readily cut with snips if the tile has been heated. The tile is usually heated with a small portable propane torch.*

11. *To cut tiles that must fit into irregular spaces, make a template of the shape into which the tile must fit, using heavy paper or cardboard. Then, transfer the pattern from the template to the tile to be cut.*

12. *When laying tile in a bathroom, ''pull'' the toilet, lay the tile, and then reset the toilet. When resetting the toilet, always use a new*

wax seal so that the joint between the toilet and the drain will be secure and watertight.

13. *Cove base tile can be installed after the floor tile has been set. A special adhesive is used to fasten the base to the wall. Cut the base to fit, check the fit, and apply the adhesive to the cove base and press it into place.*

14. *Check the completed job for any adhesive that may have worked out or smeared on the tile. Use an adhesive cleaner that is recommended by the tile manufacturer. This precaution is very important, since some adhesives can dissolve certain tiles.*

FACTORS TO CONSIDER

- When working with tile adhesive cleaners make sure the area is well ventilated, since the fumes may be hazardous.

- All tile, no matter what color or design, has a grain or flow of color. Lay each tile of the same color so that the grain runs in the opposite direction of the tile next to it. This pattern will allow for slight differences in the surface color of the tile and help ensure a better match.

- Self-adhering tiles are simple to install. Simply remove the paper backing, place the tile in position, and press it down. The layout and marking process is the same as that used for other types of tile.

- All tile should be kept at room temperature (at least 65 degrees) for 24 hours before and during installation. If the tiles are cold, they may break or may not develop a strong bond with the floor.

Basic Procedures for Installing Resilient Sheet Flooring

1. *Cushioned vinyl flooring comes in widths of 6, 9, and 12 feet. No seaming is required during the installation process unless the room is wider than 12 feet.*

2. *Unroll the sheet and spread it smoothly over the floor. Let the excess material turn up around the edges of the room.*

3. Press a straightedge into the flooring where it butts up against the wall. Draw a utility knife along the straightedge, cutting the flooring to size. Make sure the straightedge is parallel to the wall.

4. Trim the flooring with ⅛-inch clearance. This clearance will allow for expansion and contraction of the underlayment.

5. Install the baseboard molding to conceal the gap.

6. At thresholds, install a metal edge to protect the vinyl flooring and prevent it from curling up. Screw the metal edge to the floor but do not go through the vinyl.

7. Some jobs will require that the flooring be seamed. A seam can be created through the following steps. Lay the flooring out and determine the match and location of the sheets. To best match the seam, overlap the flooring where the seam edge will be created. Be sure to allow enough material in both the width and length of the flooring to match the pattern. After the pattern has been matched, weight or tape the matched pieces so that they will not shift. Cut the flooring to fit at the wall edges. Using a straightedge as a guide, cut through both pieces of material in the overlapped area with a sharp knife. As you cut, keep the knife vertical, since leaning the knife to either side will create a gap in the seam. Remove the waste pieces on both the top and bottom. Next, turn back one piece of flooring at the seam. Draw a pencil line on the floor along the edge of the second piece. Turn back the second piece and spread a 6-inch band of adhesive under the seam area using the pencil mark as a guide. The seam adhesive should be selected based on the recommendations of a floor dealer. (Seam adhesive is not the same as tile adhesive.) Lay the two pieces of vinyl flooring on the wet adhesive, and wipe it down with a damp cloth to ensure good contact with the adhesive. Press the seam together and remove any excess adhesive. If necessary, the seam can be rolled with a roller within thirty minutes of applying the adhesive.

Underlayment. Underlayment is sheet material that provides a smooth surface on which to install flooring materials like asphalt, vinyl, or rubber. Conventional subflooring may be too rough and uneven to take flooring material directly. Without underlayment, the subfloor will "telegraph" or show unevenness later through wear marks in the flooring.

FACTORS TO CONSIDER

- The newest vinyl flooring is very flexible and easy to work with. It conforms readily to room area and fits into corners tightly even before trimming. However, since this flooring is so easy to cut, it is also easy to cut it too short if care is not taken during the layout.

- Be sure to use seam adhesive where needed; it is clear and holds the seam together.

- When laying out the flooring, cardboard spacers can be inserted between the edge of the flooring and the wall to provide the ⅛ inch clearance. After layout and cutting pull out and throw away the cardboard.

- The floor molding is installed using the same procedures as those outlined in the trim and molding section.

Basic Procedures for Installing Underlayment

The two materials used for underlayment are particleboard and hardboard. The 4 by 8 sheets come in a variety of thicknesses, but the most commonly used is ¼ inch. Either type of material is appropriate; both will bridge gaps, cups, and cracks.

1. *If the subflooring has high spots, sand them down. If there are low spots, fill them with crack filler and let it harden.*

2. *Start the underlayment in one corner and fasten each panel securely before laying the next.*

3. *Follow a nailing pattern in which the nails are driven approximately 6 inches apart and ½ inch to ¾ inch from each edge. Some underlayment has a nailing pattern printed on the surface. Use ring groove or cement coated nails.*

4. *Allow at least a ⅛- to ⅜-inch space along the walls or any other vertical surfaces.*

5. *The panels can be nailed or stapled to the subfloor. If staples are used, they must be at least ⅞ inch and spaced 4 inches apart. When using nails or staples, be sure that the fasteners are set flush with the surface.*

6. *Underlayment can also be glued to the subfloor. The adhesive used is the same as that recommended for attaching subflooring to floor joists. Using an adhesive prevents the possibility of nails or staples popping up under resilient floors.*

7. *Stagger the joints of the underlayment panels. The direction of the continuous joints should be at right angles to those in the subfloor.*

WOOD FLOORING

Wood flooring provides an attractive floor covering that lends a feeling of warmth and comfort to a room. It also provides a surface that is easy to walk on, since it does not cause as much stress on a person's back and legs as other hard materials such as concrete. Wood flooring comes in a number of different forms:

• Strip flooring consists of long narrow pieces or strips with tongue-and-groove joints along the sides and often along the ends. It is also called matched flooring.

• Plank flooring consists of wider boards than strip flooring and has tongue-and-groove joints along the sides and ends. Plank floors are usually laid in random widths. The pieces are bored and plugged to simulate the wooden pegs originally used to fasten them in place.

• Parquet flooring consists of short, narrow boards cut to form patterns or mosaics.

• Wood-block flooring consists of pieces of wood cut to lengths of 2 to 4 inches, forming blocks that are set with the ends of the grain exposed to wear.

• Fabricated wood-block flooring consists of small square or rectangular blocks formed by fastening short pieces of strip flooring together. Tongue-and-groove joints are on all sides.

Wood flooring comes in hardwoods, such as walnut, birch, beech, maple, and white and red oak. The oaks may be plain-sawed or quarter-sawed (also called vertical grain or edge grain). Wood flooring also comes in softwoods, such as fir, yellow pine, white pine, and cypress. In softwoods, only the quarter-sawed, or vertical-grain, woods should be used.

The grades of wood flooring are determined by the standards set by the U.S. Bureau of Standards and by trade organizations such as the National Oak Flooring Manufacturers' Association. The grades for maple, beech, and birch flooring are first, second, and third. Oak flooring comes in five grades: clear, sap clear, select, No. 1, and No. 2 common. These grades are based on the amount of streaks, color, pinworm holes, and sapwood in the flooring.

The thickness of wood flooring can be ⅜, ½, or ¾ inch. Widths range from 1½ to 3¼ inches. For assembled squares, the sizes are in multiples of the widths of the strips from which they are made. Thus the sizes are 6¾ by 6¾ inches, 9 by 9 inches, or 11¼ by 11¼ inches.

Basic Procedures for Installing Wood Flooring

1. *The wood flooring that is to be installed should be delivered to the site four to five days before installation and piled loosely throughout the structure. This will allow the flooring to condition itself and match the moisture content (MC) of the structure. The inside temperature of the structure should be at least 70 degrees to allow for easy installation.*

2. *Be sure that the subfloor is level and clean. Cover the subfloor with a layer of 15-pound asphalt-saturated building felt that is lapped 3 inches at the seams. (See Figure 4-35.)*

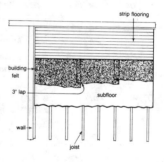

(Figure 4-35)   Subflooring

3. *Set the first course with a chalk line. The flooring is normally laid along the length of the room. The first course must be perfectly aligned, or each succeeding course will be progressively out of line and the finished floor will not be square with the room.*

4. *Leave a gap of ¼ inch between the first course and the wall. This gap will be covered by the baseboard molding. The first course is*

nailed through the center of the strip with face nails that are set below the surface; the holes above the nails are filled in later. Before setting the face nails, drill pilot holes that are slightly smaller in diameter than the nails to prevent splitting the strip.

5. *The succeeding courses are blind nailed with the nails penetrating the flooring where the tongue joins the shoulder. Blind nailing is*

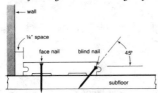

(Figure 4-36)   Nailing a subfloor

when the nails are driven at a 45-degree angle into the subfloor and thus are not visible when the floor is completed. (See Figure 4-36.) A nailset is used to finish driving the nails in so that the edge of the strips will not be damaged. Use 2- to 2½-inch cut steel flooring nails, depending on the thickness of the finish floor. A portable nailer can be used for nailing. The nailer fits over the tongue and aligns the nail in place; then the rubber head is hit with a mallet driving the nail in place. Power nailers can be used if care is taken to properly set alignment and pressure.

6. *Each course should fit tightly against the preceding one. When it is necessary to drive a flooring strip into position use a piece of scrap stock to prevent marring the flooring.*

7. *After cutting the pieces for the next course, fit them to the already installed ones to see where the joints will fall and so that the color and grain of the wood can be matched. Successive courses should not have the joints any closer than 6 inches to each other. Pieces cut from the end of a course should be carried back to the opposite wall to start the next course, since the leading edges of these pieces have no tongues or grooves.*

8. *Flooring strips should be run uninterrupted through doorways and into adjoining rooms. When there is a projection around which the flooring must be installed, take care during the layout to ensure proper fit.*

9. *To work around a projection, lay the main area of the flooring to a point even with the projection, then extend the next course all the way across the room. Set the extended strip to a chalk line and drive* *a face nail to set the strip in place. Form a tongue on the grooved edge by inserting a hardwood spline. Next, install the flooring in both directions from this strip.*

10. *Wood flooring can be laid over concrete if a nailing surface has been provided. The nailing surface is composed of 2 × 4 screeds (short lengths to which flooring can be nailed) set in mastic. The screeds are random lengths—18 to 48 inches long—and are set in mastic that is about ¼ inch thick. Along the walls, the screeds are 2 × 6 or 2 × 8 pieces. The screeds are set approximately 24 inches apart and overlap each other by about three inches. Another system of laying strip floors over concrete consists of a double layer of 1 × 2 wood screeds nailed together, with a moisture barrier of 4-mil polyethylene film placed between them. (See Figure 4-37.) This method is approved by the Federal Housing Authority (FHA). Nailable concrete may be used on some sites, but the results are not always the best. The nails have a tendency to pull out of improperly mixed concrete.*

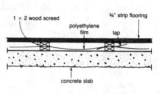

(Figure 4-37)   Flooring screeds

11. *Parquet flooring is installed in blocks that are usually in squares of 6, 8, or 12 inches. The blocks are made of several short lengths of flooring that have been glued together. Laminated blocks are made of several layers of hardwood bound together with moisture-resistant glue. Parquet flooring uses various patterns pieced to create larger designs and can be installed almost anywhere. If the flooring is to be installed at or below grade on concrete, put down a polyethylene film as a moisture barrier. Do not lay down a floor over a subfloor subject to moisture, as the blocks will warp and pull away from each other.*

12. *Parquet blocks may be laid either square or diagonally. It is best to place full-sized blocks in doorways, where traffic is concentrated. This can be done by starting the layout from the center of the wall where the door is located and laying loose blocks to a point about 4 feet into the room. Snap a chalk line from this center point to the opposite wall. Snap a second chalk line to mark the center of the*

first line. The first parquet is placed in the right angle made by the intersection of these lines at the room's center point; subsequent blocks are laid in a pyramid sequence, back towards the wall with the door. Once this first area is complete, lay the rest of the blocks.

FACTORS TO CONSIDER

- Most flooring comes prefinished and requires the utmost care to ensure that the surface is not marred. Hammer marks must be avoided, since they are very difficult to remove. Installing refinished flooring is the last activity in the construction sequence. All of the interior trim must be installed first. Protective mats must be laid over the flooring, and shoes should be checked for embedded hardware that might scar the surface.

- Moisture is an enemy of wood flooring, so every precaution must be taken to ensure that the floor is protected. If there is any concern about moisture, lay polyethylene film before installing the flooring.

- If unfinished flooring is installed, follow standard wood finishing procedures.

FLOOR FRAMING

Floor framing uses posts, beams (or girders), sills, joists, bridging, and subfloors. The specific combination of these parts depends on the design and location of the structure and the local building codes. In some parts of the country, wind, earthquakes, or climate may dictate the kind of framing used.

There are two basic types of floor framing: (1) platform, or western, framing and (2) balloon framing, which is seldom used in modern construction.

In platform framing, the structure is framed on top of the foundation walls. The floor joists and subfloor are put in place and provide a platform upon which the wall sections can be assembled and raised. The wall sections support the framing for the second story, which is also a platform. Each floor is formed separately.

Platform framing allows for the even settling of the structure and accommodates shrinkage in the materials. This type of framing also helps prevent fires from spreading in a horizontal direction.

In balloon framing, the studs are erected on the sill and run to the rafter plate. Ends of the second floor joists are supported on a ribbon. They are spiked to the studs as well. Fire-stops must be added to the space between the studs.

The design of balloon framing reduces shrinkage because the amount of cross-sectional lumber is low. Wood shrinks across its width; practically no shrinkage occurs lengthwise. Balloon framing adapts very well to construction in which veneer or stucco is used on the outside walls.

In order to support the floor joists and to span wide areas such as basements, beams are used. Beams can be either solid or box and can be built up with two or more pieces of 2-inch wide lumber and nailed together with 20d nails. The end joints are staggered, but they usually join over a post.

Short beams can be butt-joined by using special metal connectors. At least 4 inches of the beam end should rest on masonry walls. The top of the beam should be flush with the sill plate unless notched joists are set on ledger strips. Beams are spaced to allow utility lines to run between them.

In order to determine the exact size of the spanning beam, the following formula is used. Find the distance between girder supports, and then find the girder load width. (A girder must carry the weight of the floors on each side to the midpoint of the joists that rest on it.) Determine the total floor load carried by the joists and bearing partitions to girder. This is done by figuring the sum of loads per square foot. Local building codes will give the specific requirements on figuring sum loads. Be sure to check on whether the roof loads are included; in many cases they are not, since this weight is carried on the outside walls.

Check the plans to see whether braces or partitions are to be placed under the rafters; if they are, then the roof load must be figured in, since the braces and partitions transfer part of the roof load to the floor load. To find the total load on the girder, multiply the girder span by the girder load width by the total floor load (see the Average Load chart in the Troubleshooting section). Next, select the proper size of spanning girder that is in compliance with the local building code.

On many construction projects steel beams are used instead of wood girders. The size of the steel beams for the specific job is calculated in the same way as for the wood girders. The W (wide-flange) beam is the type usually used in residential construction. In commercial construction, the beams are usually custom ordered to meet a specific spanning need and are detailed in the architectural plans. Steel beams vary in depth, width of flange, and weight. After figuring the floor load, consult with a steel supplier about the specific beam needed for the span.

The girders or beams are supported by posts or columns. These posts must set on a firm base or foundation in order to adequately support the beams that in turn support the floor joists. The base or footings are set above the floor level to create a pedestal upon which the post can rest. To make sure that the post does not slide off of the pedestal, a ½-inch reinforcing rod or bolt should be embedded in the footing before the concrete sets. The rod should project about 3 inches above the concrete. A hole is drilled in the bottom of the post, and the post is placed over the rod. Alternatively, a manufactured post anchor can be used. The anchor is set in the concrete and slides onto the bottom of the post. The anchor holds the post off of the floor and protects the wood from moisture. The anchor has an adjustment that can be set for plumb when installing the post.

Wood columns that are going to support a steel beam should have a metal cap on the top. The cap provides an even support surface between the column and the beam and prevents the end grain of the column from being crushed.

The width of the wood posts should be the same as the width of the beams or girders that they support. This rule holds true for beams or girders that are not longer than 9 feet or smaller than 6 inches. For example, an 8-inch wide beam should have a post that is 6 inches by 8 inches or 8 inches by 8 inches. The posts can be solid wood or built-up by spiking 2 inch lumber together. If posts are made from 2× material, the pieces should be carefully selected to make sure they are free of defects that might cause one of them to buckle and put extra pressure on the remaining members.

The most commonly used posts are made of steel. The steel post is capped with a steel plate designed to provide a balanced load-bearing surface. Steel posts have a threaded area inside the top end so that they can be adjusted to the exact length needed; when changes occur in the structure as a result of the building settling. The posts can accommodate these changes.

FLOOR TRUSSES

Floor trusses are fabricated with lumber chords and patented galvanized steel webs to span long runs firmly and with a minimum of depth. The trusses' open lattice design results in light-weight and easy assembly. The lattice allows for easy installation of plumbing and electrical lines, while reducing transmission of sound through floor and ceiling assemblies.

Floor trusses are widely used in construction today. They are easy to install, since they come to the job site ready to be set in place. They are designed with the aid of computers, and the maximum load requirements are calculated so as to use a minimum of materials. These calculations take into account tension and strength in the placement of the steel webbing and ribbing. Trusses should be considered for all jobs, based on availability, cost, and local building codes.

JOISTS

The framing members that carry the weight of the floor between the sills and girders are called floor joists. In residential construction, wood joists are used, and they are usually 2 inches thick and are placed on edge. Commercial construction normally uses concrete reinforced with steel bars. Joists are usually spaced 16 inches O.C. In some cases, the spacing may be 12, 20, or 24 inches O.C. The plans will detail the exact width to be used, or the width can be calculated based on the floor load. Floor loads are calculated based on an average of 50 pounds per square foot—10 pounds dead load and 40 pounds live load. Joists must carry the loads placed on them and be stiff enough to prevent undue bending or vibration. Building codes usually specify that the deflection (bending downward at the center) must not exceed 1/360 of the span with a normal live load. This would equal ½ inch for a 15-foot span.

Basic Procedures for Laying and Constructing Floor Joists

1. *Study the plans carefully. Know the direction in which the joists must run and the location of supports such as posts, columns, and partitions.*

2. The joists can be laid out directly on the sill, if balloon framing is being used. With platform framing, the joist locations are laid out on the joist header rather than on the sill.

3. The joists are laid out by marking a line on the sill and then placing an X near it to indicate the position of the joist. A marking rod can be used to mark the location for framing members. A marking rod is a 1-inch thick piece of lumber approximately eight feet long that has the location of the framing members laid out on it. The marking rod is placed along the sill. The position of the joists is then transferred to the sill, using the rod as a guide.

4. When the joists are lapped at the girder, the X for their location is marked on the other side of the layout line for the opposite wall. In this case, the spacing between the 16- or 24-inch on center (O.C.) stringer will be different from the regular spacing by 1½ inches, due to the offset created by the overlapping of the joists.

5. Double the joists where extra loads must be supported. If a partition is to run parallel to the joists, a double joist must be placed underneath the space where it will be located. If pipes are to run through the partition, it is generally made wide enough to accommodate them, and the joists are set to match the partition width. Trimmer joists must be used around openings in the floor frame for stairways, chimneys, and fireplaces. These joists are doubled and are used to support the headers that carry the tail (short) joists. Select straight lumber for the header joists and lay out the standard spacing along the entire length. Mark the locations for any doubled or trimmer joists that will be required along openings. The regular joists that become tail joists are marked with a T instead of an X; the T will remind you where the tail joists are to be set.

6. When the header joists have been laid out, toenail them to the sill.

7. The full-length joists should be positioned with the crown (the edge that has the longest grain lines) turned up.

8. Push up the joist tight against the header and alongside the layout mark. Nail the joist to the header using three 16d nails.

9. Nail the joists along the opposite wall. If the joists butt at the girder and do not overlap, they should be joined with a scarf or metal fastener. If the joists overlap, no metal fasteners are needed. They can simply be nailed together.

10. *Some carpenters prefer to nail the joists to the headers before fastening the headers to the sill. They then slide the headers around until they are perfectly aligned on the foundation or with the sill.*

11. *The doubled joists should be nailed together using 12d or 16d nails spaced about 1 foot apart along the top and bottom edge. Drive the first nails straight through the lumber to draw the stock together. These nails are then clinched over. The rest of the nails should be driven in at a slight angle so they do not protrude through the other side.*

12. *When framing openings, such as stairways, make sure you have a solid surface on which to work. Place sheets of plywood across the joists and work on these.*

13. *To frame openings, lay out the length of the headers using the main header joist as a pattern. Cut the header and tail joists to length. Be sure that the cuts are square. Much support strength is lost if the surfaces do not match up and fit tightly with each other.*

14. *Nail the headers in place and then nail the tail joists to them. To help hold the headers in place for nailing, a tail joist can be temporarily nailed to each trimmer. The tail joist gives the support needed to accurately nail the headers in place.*

15. *Nail the tail joists in place.*

16. *Nail the second headers in place.*

17. *Nail the second trimmer joists in place.*

18. *Proceed with the installation of the regular floor joists. They are put in after the opening is framed because of the need for nailing clearance.*

(Figure 4-38) Joist hanger

19. *When nailing the joists in place, drive three 16d nails into the headers and joists.*
    Where double trimmers or headers are used, continue to use three nails per joist but use a nailing pattern in which the nails are slightly staggered above and below a horizontal line.

20. *Metal framing anchors, or joist hangers, are used extensively in*

construction. They are made from 18-gauge zinc-coated steel, and come in a variety of sizes and shapes. (See Figure 4-38.) These hangers come with special nails to maintain holding power. The hangers provide for two-way nailing, as opposed to simple toenailing, in joists where it is not possible to nail through the headers. It is recommended that framing anchors be used to support tail joists over 12 feet long.

FACTORS TO CONSIDER

- Use the correct size nails to frame floors. Also, follow a logical nailing pattern so that the joists supply maximum weight support—eliminating any worry about the floor giving way or moving back and forth. A properly balanced nailing pattern will let the floor support a concentrated load of 300 pounds at any given point.

- The headers and joists must fit tightly against each other. If they do not, they can work back and forth which may cause the nails to break and the floor to drop or sag.

- Power nailers may be used to save time and effort, but the gun must be set correctly for sufficient penetration of the nails into the wood.

Basic Procedures for Framing over Girders and Beams

1. When framing joists over steel beams, place a 2 inch wood pad on top of the beam. This brings the joists up even with the sill. (See Figure 4-39.) When a wooden girder is used, it is set so that its top is even with the sill.

2. The joists can be notched to move the ceiling height under them. Part of the floor load from the joists is then carried by a ledger (a nailing board that has been fastened to the foundation wall). Be sure to align the ledger with the bottom of the girder and to cut the notches accurately so that the weight of each joist rests on the ledger, not on the notch.

3. If the joists and the girders must be even on the underside, use joist and beam hangers. These hangers can be adjusted as needed to

maintain an even surface on the underside. This is important when a basement ceiling is going to be installed directly below the floor joists.

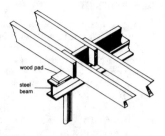

(Figure 4-39) Framing joists

4. Joists may be framed to steel beams at various levels with special hangers. These are used in essentially the same way as the hangers for wood girders. A key factor is to allow for shrinkage in the joists, because the size of the steel beams will remain unchanged. A clearance of ⅜ inch above the top flange of the steel girder or beam is usually sufficient.

5. Joists can be notched to fit on the bottom flange of steel girders if a wide-flanged beam is used. Do not rest the joists on S-beam girders, because the bottom flange does not provide enough surface upon which to rest the joist and its load.

6. When the joists have been secured in place, bridging is added to lend additional support to the joists. Bridging is used to hold joists in a vertical position and to transfer the load from one joist to the next. Bridging can be cut on the job, but most jobs use precut wood bridging or steel bridging.

7. Bridging is put in place after a chalk line has been snapped across the tops of the joists to show where the bridging is to be set. Start two 8d nails into the ends of all the bridging and then attach a piece to each side of every joist. Alternate the position of the two pieces, first on one side of the chalk line and then on the other side at the next joist.

8. The bottom of the bridging is not nailed into place until the subflooring is completed or the underside of the floor is enclosed.

9. If steel bridging is used, no nails are required. The bridging is set in place, and the claws on each end are driven into place.

10. In some cases, where the joists have to be held tightly in a vertical position, solid bridging is used. This bridging is solid dimension lumber cut to fit between the joists and nailed into place.

SILL CONSTRUCTION

Sill construction follows installation of the girders and beams. The sill is attached to the foundation wall and serves as a base for other framing work. The sill usually consists of 2 x 6 inch lumber and is fastened to the foundation with anchor bolts or straps. Local building codes specify the size and spacing of the anchors.

Before the sill is set in place, termite shields and sill sealers are installed to keep out insects and prevent drafts. A termite shield of not less than 26 gauge sheet metal, is extended over the foundation. Also, the wood sill should be at least 8 inches above the ground. In areas where there are heavy infestations of termites, additional measures such as poisoning the soil around the building or treating the wood with chemicals may be necessary.

Sill sealers are sold in 50-foot rolls and used to prevent drafts between the foundation and the sill. There are a number of different materials from which sealers are made, so one should check the building codes for specific requirements and a building supplies outlet for the most suitable material for the local area.

Basic Procedures for Installing Sills

1. *If bolts are being used to anchor the sill, remove the nuts and washers. If straps are being used, check for clearance so the sill can be slipped into place.*
2. *Lay out the sill for position on the foundation. Remember to set the sill back from the outside wall a distance equal to the thickness of the sheathing.*
3. *Draw lines across the sill on each side of the bolts. Measure the distance from the center of the bolt to the outside of the foundation and subtract the thickness of the sheathing. Use this distance to locate the bolt holes. Make separate measurements for each anchor bolt, since the foundation will be uneven and the bolt setting not consistent. (See Figure 4-40.)*

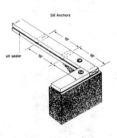

(Figure 4-40)   Installing a sill

4. *To ensure that the sill is set perfectly straight on the foundation, a chalk line can be snapped along the top of the foundation where the sill is to be set. A straight sill is necessary since all other framing stems from it. If the chalk line method is used, the variations between the outside surface of the sheathing and foundation wall can be shimmed when siding is installed.*

5. *After the locations for the holes have been marked, drill the holes. Make them at least ¼ inch larger than the bolts. This will allow some leeway in the final fitting.*

6. *Make trial fits until the sill is properly positioned.*

7. *Install the sill sealer.*

8. *Install the termite shield.*

9. *Place the sill on the foundation and tighten the bolts. If straps are being used, bend them over the top of the sill. If necessary, nail the straps in place.*

10. *Check the distance from the bottom of the sill to the top of the foundation wall. Also check to see that the sill is level. If there are gaps between the bottom of the sill and the top of the foundation, fill in this area with grout or mortar.*

SPECIAL FRAMING PROBLEMS

Special framing problems arise when a particular design is used within the structure. For example, one common design in bilevel and trilevel houses has an upper structure that overhangs the first level or basement. Framing such an overhang with joists that run perpendicular to the walls is very simple. You use longer joists separated by spacing blocks that keep the joists upright and rigid. This sort of blocking also closes up the space between the interior of the foundation and the on-running joists.

Framing in with cantilevered joists is necessary for joists that run parallel to the wall. The cantilevered joists should extend toward the main part of the structure at least twice as far as they stick out over the supporting wall. Framing anchors are the easiest way to do this framing project. If anchors are not used, then a ledger strip must be placed on the double inside header. Locate the ledger on the top side of the double header, as the stress and weight are borne by this upward side. Blocking, to keep the joists aligned and upright, must be installed where the centilevered members go from the supporting wall outside.

If concrete is used to provide a base for slate, stone, or tile, the floor frame must be lowered. If the area is small, such as for an entry way, use smaller dimension double floor joists, set closer together. Use steel or wood girders and posts for larger areas.

Bathrooms require extra framing. A tile floor will add about 30 pounds per square foot of floor; fixtures will add 10 to 20 pounds, for a total of 40 to 50 pounds of dead load per square foot. Also, in many cases the joists must be cut to bring in water service and waste pipes, and special precautions must be taken to add support.

To maintain the maximum strength of joists, holes for plumbing should be limited, if possible, to one fourth of the total width or depth of the joist. If this is done, no material reduction in strength will result. If cuts are made in the joists, make the cuts from the top. Top cuts cause a loss of compression; bottom cuts cause a loss of tension, which can result in the joist collapsing. Also, joists must be reinforced with headers and trimmers, or by adding extra joists where large cuts are made. Another way of maintaining joist strength in large cuts is to make the cut in two stages: In the first stage, make a cut to a depth of about 2 inches; in the second stage, make the cut needed to accommodate the plumbing pipes. The first-stage cut is filled in with 2 by standard dimension lumber, reinstating the strength of the joist while leaving an area for the pipes.

When cutting joists, the following information should be considered. When a 2 × 10 joist is cut to a depth of 2 inches, its strength is reduced to that of a 2 × 8. The weakest point of a joist is the center of the span, so try to make the cuts or holes as far away from the center of the span as possible. If center-span cuts are necessary, make sure the span is reinforced with headers or trimmers.

SUBFLOORS

The final step in completing the floor frame is the installation of subflooring. Subflooring adds ridgidity to the structure, provides a base for finish flooring material, and provides a surface upon which additional construction can take place, such as wall layout and framing.

There are several different materials for subflooring; ship-lap, common boards, and tongue and groove boards have been used over the years. These materials are nailed to each joist using 8d nails. If the boards are less than 6 inches wide, two nails are used; if the boards are wider than 6 inches, three nails are used. The boards are set at 45 degrees to the foundation and are butted tightly together without any cracks. If the subfloor is likely to be exposed to an accumulation of water during the construction process, then cracks should be left to allow for drainage. Subflooring boards are not used very much today because of the development and improvement of sheet subflooring materials.

The most commonly used sheet material for subflooring is tongue-and-groove plywood. It provides a smooth base and gives horizontal strength to the building. It installs quickly and is relatively squeak-free. Most builders use ⅝- or ¾-inch thick stock. This thickness ensures adequate floor strength for nearly every situation. The sheets are installed perpendicular to the joists, with the joints being staggered in successive courses. The sheets are glued to the joists, using a standard waterproof adhesive, and nailed using 8d nails spaced 6 inches along the edges and 10 inches along intermediate members. (See Figure 4-41.) This method of installation ensures squeak-free floors, eliminates nail popping, and reduces labor costs.

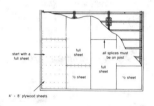

(Figure 4-41)   Subflooring

There are other sheet materials that are available for use as subflooring. These include composite board, waferboard, structural particleboard, and oriented strand board (OSB). They all have been approved and rated by the American Plywood Association as being suitable for use as subflooring and are installed in the same manner as plywood.

Basic Procedures for Installing Sheet Subflooring

1. *The long dimension of the sheet should run perpendicular to the joists. Joints should be staggered in successive courses.*

2. *Use 8d nails positioned 6 inches apart along the edges and 10 inches apart along intermediate members.*

3. *When using glue, put a ¼ inch bead along the center of the joists. Spread enough glue to lay one or two panels. Don't do more than this, because the glue will set and get in the way of the work area.*

4. *Where two panels butt together, lay two beads of glue, to ensure that both panels get adequate glue on their surfaces.*

5. *If the sheets are tongue and groove, lay a ⅛ inch bead of glue. This smaller bead will avoid having excess glue squeeze through the seams.*

6. *Tongue-and-groove sheets must have an ⅛ inch space all around them. Most of the grooves are set so that this spacing is automatically achieved when the sheets come together.*

7. *Tongue-and-groove sheets can be driven together with a maul (a sledge-type hammer) and protective wood strip. When driving the sheets in place, drive only from the groove side to prevent damaging the sheet.*

8. *Complete all nailing before the glue sets.*

FOOTINGS AND FOUNDATIONS

No structure is any stronger than its foundation; this is true whether the structure is a small shed or a multistory office building. This section starts with the basic layout of footing and continues through to the completion of the foundation.

A first step in the construction process is analyzing subsoil and drainage conditions, water-table depth, soil compaction, and local building codes, as well as securing the necessary building permits.

FOOTINGS

Footings distribute the building loads over a sufficient area of soil to secure adequate bearing capacity. They are classified in the way they receive the loads.

CANTILEVERED FOOTINGS

Cantilevered footings are used to solve the problem of providing proper support for the foundation while not going over the property line. This footing system ties the footings for the outside row of columns to those of the first inside row in such a way that any rotational (turning) tendency is neutralized.

COLUMN FOOTINGS

Column footings are the most commonly used isolated, or independent, footings. For light loads, the simple type of column footings are used. The columns are set at a constant depth with two-way reinforcement. More complicated column footings are used in large construction projects where the columns are set on bedrock. Simple column footings are used over a square pad of concrete, with or without steel reinforcement. The footing accepts the concentrated load placed on it from above by a building column and spreads it across a large enough area of soil so as not to exceed the allowable amount of stress to the soil.

COMBINED FOOTINGS

Combined footings are a combination of column and wall footings. This type of footing is most frequently used to support walls and columns near property lines. The wall column footings combined with interior column footings provide the support needed without projecting over the property line. This combination of footings also eliminates the rotational tendency of the structure.

INDEPENDENT FOOTINGS

Independent footings are those footings required as a result of some special building requirement and are independent of the other footings within the structure. A common example of this is a masonry chimney. Chimney footings should have a minimum projection of 4 inches on each side. For a two-story house, the chimney footings should have a minimum thickness of 12 inches and a minimum projection of 6 inches on each side. Independent footings should be laid out and poured at the same time as the other footings.

STEPPED FOOTINGS

Stepped footings are used when a building is constructed on a slope and support must be given at all points along the slope. The footings are formed and cast like wall footings, but they are set in steps. The steps are sized based on the degree of the slope and the size of the structure. If soil conditions are unstable or if the building is in an earthquake zone, the column footings on steep slopes are tied together with reinforced concrete tie beams.

WALL FOOTINGS

Wall footings support light loads and are constructed of plain concrete that does not project more than about 6 inches beyond the edges of the wall. The minimum footing depth is usually equal to twice the projection. Minimum code requirements state that light longitudinal reinforcement must be embedded in simple concrete wall footings. This is required to minimize shrinkage and cracks caused by temperature extremes and to bridge over soft foundation material. These footings are generally known as T-footings and are poured at a constant depth around the perimeter of the foundation. Another common wall footing is the square footing, which is similar to the T-footing. Wall footings are used with three common substructures: slab on grade, crawl space, and basement.

Basic Procedures for Building Forms for Footings

1. *After the excavation is completed, check the batter boards to see whether they are still correct. Often, they are knocked around during excavation, so they must be reset or at least checked for accuracy.*

2. *Lay out the lines on the batter boards; where the two lines intersect, drop a plumb bob to the bottom of the excavation. This locates the building corner. (See Figure 4-42.)*

3. *Drive a corner stake and drive a nail in the top of the stake at the exact point the plumb bob touches it to establish points at the corners of the foundation walls.*

4. *Set up a builder's level at a central point in the excavation and drive a number of grade stakes (level with the top of the footing) along the footing line.*

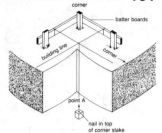

5. *Drive stakes at the approximate points where column footings are required.*

6. *Drive corner stakes to the exact height of the top of the footing. Connect the corner*

(Figure 4-42)   Laying out a foundation wall

stakes with lines tied to nails in the top of the stakes.

7. *Using these lines as guides construct the outside of the footing form. (See Figure 4-43.)*

8. *The form boards are located outside the building lines by a distance equal to the footing extension (usually 4 inches for an 8-inch foundation, 6 inches for a 12-inch foundation wall). The top edge of the form boards must be level with the grade stakes.*

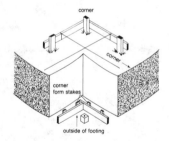

9. *Transfer the measurement from the grade stakes to the form. Use a carpenter's level to do this.*

(Figure 4-43)   Footing frame

10. *Forms are generally constructed of 1-inch boards and should be supported with stakes placed 2 to 3 feet apart. If 2-inch material is used, the stakes can be placed farther apart. Once the outside forms are in place, it is relatively easy to set the inside sections.*

11. *Cut spacers to length for the inside form. This will save time, since the inside form is set next to the end of the spacer and is already set at the proper distance inside of the outer form. (See Figure 4-44.)*

12. *The forms may need to be braced, especially if 1-inch stock is used.*

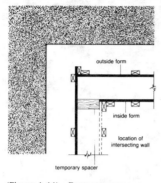

outside form

inside form

location of
intersecting wall

temporary spacer

(Figure 4-44)   Frame spacers

Attach the braces to the top of a form stake and to the bottom of a brace stake located about a foot away.

13. *Additional form work will be required with stepped footings. Vertical blocking must be nailed to the form to contain the concrete until it sets.*

14. *For column footings, the size of the form boards depends on the column being created. The top of the form boards must be level with the grade stakes. Often the hole for the column footing must be dug out. Make sure all loose dirt and debris have been removed from the hole so that no pockets are created in the concrete.*

15. *When forming up for footings, use duplex (double-headed) nails. They hold well and are easily removed when pulling the forms.*

16. *When nailing the forms, be sure to nail through the stakes into the form boards. This procedure will make the removal of the forms much easier later on.*

17. *Check all of the forms for accuracy and sturdiness. Remove the lines, line stakes, and grade stakes. The forms are now ready for pouring.*

FOUNDATIONS

The foundation transfers the weight of a structure to the earth along the base of its walls or its columns. The foundation is also used as a building base, such as in slab construction, or as a part of a storage or living area, such as a basement.

BASEMENT FLOORS

Basement floors are similar to slab on grade construction in that the subgrade must be firm and completely free of sod, roots, and debris. If needed, coarse fill can be used. This fill is placed over the finished subgrade and is 4 inches thick. It must be well compacted. All service lines must be laid out and installed before the floor is poured.

A common problem with basements is water seepage. This problem must be planned for when preparing to pour the basement floor. If groundwater is a concern, drain tile should be installed around the outside of the basement walls. Drain tile laid around the inside of the basement walls and set to drain into a sump pit is another way of solving water problems. If a sump pit is put in, check the plumbing plans on how the pit is to be connected into the sewer system; be sure that power is run to the pit for the sump pump. Basement floors are generally poured later in the construction process, after the basement walls have been poured. When pouring basement floors, there is no need to build forms, since the basement walls serve as the forms. Basement floors are given a smooth finish that can be sealed or prepared for a surface flooring material, such as tile or carpeting.

CONCRETE

Concrete is a mixture of portland cement and water, combined with aggregates (commonly sand, gravel, and crushed rock). The cement and water create a paste that holds the aggregates together in a rocklike mass. Cement is manufactured from limestone, silica, alumina, and lime. Through processing, these raw materials are turned into cement with characteristics that vary depending on the material composition and manufacturing process used.

There are five types of portland cement used in construction. The most common is Type I, which is used in constructing columns, beams, and floor slabs and for other general construction purposes.

Concrete is usually delivered to the job site premixed. It is dumped on the site where the concrete will be used, or it is moved from the truck to the finish site by power buggy or wheelbarrow.

The consistency determines the workability of the concrete. The test for consistency of concrete is called slump testing. Slump testing is performed by engineers or technicians and is not required on the construction job. An experienced concrete worker can determine whether the mix is correct by observing the texture. Water can be added to make the concrete workable. The concrete supplier can assist in getting the right mix for the job. Working with concrete requires skill to develop the smooth hard finish desired on basement floors or slab construction. Guidelines for working with concrete are covered in the section on sidewalks and driveways.

When using concrete for footings, no finish is required, but the footing surface must be even and smooth. When pouring basement walls, a power vibrator must be used to ensure that all the air bubbles are removed from the concrete, especially at the bottom of the forms. The vibrator is inserted into the concrete every few feet along the walls. This settles the concrete by removing excess air and provides a solid wall and smooth finish.

CONCRETE BLOCKS

Concrete blocks are used for foundation walls and other masonry construction, although they are used less today because of the refinement of poured concrete walls. Standard concrete blocks are made from portland cement and natural or manufactured aggregates. Among the aggregates are volcanic cinders, pumice, and foundry slag. The blocks weigh between 25 and 35 pounds. Blocks should comply with specifications provided by the American Society for Testing Materials. Concrete blocks have a lower U factor (measurement of heat flow or heat transmission through materials) than poured concrete walls.

Blocks are classified as solid or hollow. A solid unit is one in which the core, or hollow, area is 25 percent or less of the total cross-sectional area. Blocks are available in widths of 4, 6, 8, 10, and 12 inches and in heights of 4 and 8 inches. Sizes are actually $\frac{3}{8}$ inch shorter than their nominal (name) dimensions to allow for the mortar joint. For example, the $8 \times 8 \times 16$ inch block is actually $7\frac{5}{8} \times 7\frac{5}{8} \times 15\frac{5}{8}$ inches. With a standard $\frac{3}{8}$-inch mortar joint, inches.

When using concrete blocks to build foundation walls, openings are spanned by using lintels. Lintels are made from precast concrete units that include metal reinforcing bars or steel angle irons. Lintels are made to fit window, door, or crawl-space openings. They are set in place, and the blocks are laid around them.

CONSTRUCTION PROCEDURES

CLEARING THE SITE

Clearing the site is the first step in the construction process and one that must be carefully carried out. Trees or other barriers in the area where the structure will be built must be removed and the area must be graded. If the area is wooded, check with the owners to identify the trees that are to remain. Clearly mark these. Do not cut trenches too close to the bases of these trees or put fill around the trunks. Trees cannot tolerate more than a foot of fill around their trunks.

Grade leveling is the process by which the surface level of the construction site is measured at several points; the differences are averaged and all points are brought to the same level. A leveling instrument is used to calculate the difference in elevation. If the building site slopes, the instrument is set up between the points. The reading is taken with the rod in one position, and then the instrument is carefully rotated 180 degrees to get the reading at a second position. If grade stakes or batter boards are being set, the instrument should be set in a central location to allow for each corner to be focused in without moving the instrument.

EXCAVATION

The excavation of the building site is completed in stages. The first step is rough grading, the topsoil is removed and piled out of the way, so it can be used later. If the building site is relatively even and no grading is needed, the building site can be laid out with no excavation.

Excavation for a basement requires laying out grade stakes on the outer edge of the rough excavation. The lines between the batter boards are removed during excavation so they will not be in the way. For regular basement excavation, the hole should extend 2 feet beyond the building lines to allow room for form setting. The vertical views of the architectural plans provide the information needed to calculate the depth of the excavation. By using the highest elevation on the perimeter of the excavation, a control point can be established from which the depth of the excavation and the height of the foundation can be determined.

Foundations should extend about 8 inches above the finished grade. This height adequately protects the siding and framing members from moisture. The finish grading should slope away from the structure so that surface water will run away from the building.

If the foundation is going to be slab or have a crawl space, then the excavation will extend only as far as is needed to pour the footings. In climates in which frost is a factor, the footings and the foundations must be located below the frost line; otherwise, the movement of the soil can crack or break them. Local building codes specify the requirements on solving the frost-line problem.

LAYING OUT BUILDING LINES

Building lines are laid out after the site has been cleared and the lot lines have been checked. The lot lines are best checked by a

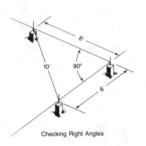

Checking Right Angles

(Figure 4-45) Layout lines

registered engineer or licensed surveyor. Either expert can also help establish building lines and grade levels. Building lines are laid out best with the use of a leveling instrument. Using the procedure outlined in the section on leveling instruments, lay out the building lines perpendicular to existing lines such as lot lines. Building lines must be square, and this can be done by using the 3-4-5 method. Any multiples of 3-4-5 can be used, such as 6-8-10, 9-12-15, or 12-16-20, to check right angles. After the corners have been formed by intersecting the outside dimensions of the foundation walls, mark the positions by driving stakes. Place tacks in the tops of the stakes where the exact center of the corner is located. (See Figure 4-45.) Recheck all building lines for accuracy. Measure the lengths and diagonals of squares and rectangles, even though they were laid out with a transit. An out-of-square foundation will cause problems all through the construction process. This is the point at which such problems must be corrected to save time, materials, and effort.

Next, batter boards are set up around the building layout stakes. The batter stakes are 2 × 4 pieces, and the ledgers are 1 × 6 or wider lumber. The batter boards are located at least 4 feet from the corners created by the building lines. (See Figure 4-46.) The ledger boards are nailed to the stakes at a convenient working height. It is best to set the ledger boards level with each other and slightly above the top of the foundation. The batter boards should be roughly level with each other. The ledger boards should extend well past each corner, so they must be of sufficient length to reach these points. Lay out the building lines by stretching string from each batter board directly over the layout stakes. A plumb bob should be used to square the lines over the layout stakes and marking tacks. Mark the top of the ledger boards where the lines cross. This can be done by cutting a slight saw kerf or driving a nail at the mark point. Pull the lines tight and fasten them. (See Figure 4-47.)

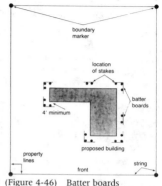

(Figure 4-46)   Batter boards

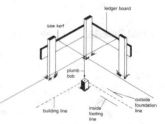

(Figure 4-47)   Tightening the lines

ERECTING WALL FORMS

Wall forms come in many different systems, but certain basic principles apply to erecting forms for all systems.

Basic Procedures for Erecting Wall Forms

1. *All wall forms must be tight, smooth, free of defects, and properly aligned.*

2. *The joints between panels must be tight. This prevents concrete from escaping. Concrete will honeycomb around joints when it flows out.*

3. *Brace the panels so that they will resist the side pressure created when the concrete flows in the forms.*

4. *For low walls, 1-inch sheathing boards or ¾-inch plywood is suitable for forms if the studs are set about 2 feet apart. For higher walls, the panels must be backed with wales (wood or metal strips used to keep the concrete from pushing the form out of shape), and the studs must be set closer together.*

5. *The pressure on the wall forms occurs as the concrete is poured and while it is in its plastic state. As the concrete hardens, it is self-supporting. So, the forms must be properly supported, based on the amount of concrete that is poured per hour, the temperature (which affects setting time), and the amount of mechanical vibration. It is always better to have too much support rather than too little. This will avoid the risk of having a wall panel give way, spilling two or three tons of concrete into the basement area.*

FORM PANELS AND HARDWARE

Foundation walls are made using prefabricated forms. These forms are generally steel panels that come in different heights and widths. Some of these panels come with designs on the inside. For example, these designs create the appearance of brick on both the outer and inner surfaces of the foundation walls. If a design is desired on only one surface the panels can be set to do this as well.

Spreaders are used to keep the panels at the desired uniform width throughout the entire foundation wall. These spreaders, or ties, are either flat metal pieces or metal rods. The flat metal pieces are ⅛-inch thick and are as long as the wall thickness, plus clearance for end clips that clip into the panel forms. The rods are set into the panels and have washers and bolts fastened to the ends to hold them in place. Once the concrete has set and the forms are pulled, the ends are snapped off the metal pieces and rods. These pieces break off flush with the concrete or slightly below the surface. If an indentation is left on the surface, it can be filled with grout.

Carpenters are sometimes called on to build their own panels. Wood panels are constructed by framing ¾-inch plywood with 2 × 4 studs to form 2 foot or 4 foot by 8 foot units. (See Figure

4-48.) Wood spacers can be used on top of the panels, and wire or metal spacers can be used inside the wall space. These wooden panels must be supported to keep them from bulging. Wall openings are planned for in wall panels by set-ins, which are metal frames in which doors or windows will be placed. Where small openings are needed, they can be made by using tubes of plastic, fiber, or

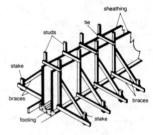

(Figure 4-48)   Forming the panels

metal. They are held in place with wood or plastic spacers that are attached to the panels. If concrete blocks are used for the foundation walls, openings are built into a wall and bridged or spanned with metal or cast concrete lintels.

FOUNDATION SYSTEM

Foundation systems must be designed to accommodate some settling. A properly designed foundation system will allow for settling so that the weight of the structure is distributed evenly and the results of the settling are negligible or at least uniform. For light construction, such as residential, the spread foundation is the most commonly used, since it transmits the load through the walls, pilasters, columns, or piers. These load-bearing devices rest on footings that provide an enlarged base. The section on footings outlines the various footings that can be selected to meet particular construction needs.

BACKFILLING

Backfilling is completed after the foundation has cured and has been waterproofed. "Backfilling" means filling the area between the foundation walls and the surrounding area. If the foundation is block construction, the walls must be supported from the inside to withstand the pressure of the fill from the outside. If tile has been installed around the outside of the foundation, care must be taken while dumping the first few loads of fill to avoid disturbing or breaking the tile. Backfilling is normally done with heavy equipment like a backhoe, so the operator of the equipment must be careful to avoid damaging the walls—especially when completing the top fill around the basement windows, stairs, and entrance platforms.

SIDEWALKS AND DRIVEWAYS

Sidewalks and driveways are put in after the finish grading is completed. If excessive fill has been used to bring the work to grade, let the fill settle before starting the concrete work. Sidewalks are generally poured directly onto the ground. Where moisture and frost are problems, a bed of coarse fill should be put down first. Forms for sidewalks are 2 × 4 lumber. The forms must be supported with stakes to insure that they remain in place and vertical.

When a sidewalk is being poured at two levels with steps between, pour the top section first, then the bottom, and then the stairs. Forms for sidewalks should slope about ¼ inch per foot during the run. This is to allow water to drain off so that it won't freeze and cause damage to the concrete.

When pouring a walk next to a building or retaining wall, do not allow the concrete to bond to the wall. Insert an expansion joint between the walk and the wall. This will allow the walk to float on the ground so that it will not crack as readily as if it were bonded to the wall. The portion of the sidewalk that will be driven over should be made 6 inches thick. Sidewalks are 4 feet wide on main walks leading to front entrances and 3 feet for secondary entrances.

Driveways should be 5 or 6 inches thick with reinforcing mesh included. A single driveway is at least 10 feet wide, and a double driveway should be a minimum of 16 feet wide. The forms are constructed of 2 × 6 lumber.

When working the concrete for sidewalks or driveways, try to get the concrete as close to its final position as possible. Once the form is filled, screed it immediately. Screeding is done by moving a straight board across the top of the forms in a seesaw motion. Always keep some concrete in front of the board. After the screeding is done, the concrete should be floated using a float made from wood or aluminum. This operation, when performed correctly, removes high spots, fills depressions, and smooths irregularities.

When the concrete has stiffened and the water sheen has disappeared, the control joints can be cut, and the edges rounded. The edges are rounded using a special edge tool that slides along the forms around the entire area that has been poured. The rounded edge on the concrete eliminates sharp corners that could break off or cause injury.

A control joint is cut into concrete to control the location of cracks in the concrete. A control joint should extend to a depth of at least one-fifth of the thickness of the concrete. A control joint is located approximately as often as the slab is wide. (For example, if the slab is three feet wide, the control joints should be cut in every three feet.) Special masonry tools can be used to cut control joints; a masonry blade and saw can also be used. If the joints are cut using a saw, they should be cut 18 to 24 hours after the concrete has been poured.

Do not overwork concrete while it is still plastic, because this will bring excess water and fine material to the surface. This will result in scaling and dusting after the concrete has cured.

STEPS AND ENTRANCE PLATFORMS

Steps and entrance platforms need to be cast as a part of the foundation walls, if possible. If this is not possible, reinforcing bars must be set into the foundation walls so that the steps or entrance platforms can be poured around them later. Some layouts include the steps and platform as one pour. When doing this, use 2-inch material on steps that are over 3 feet wide. This will prevent the risers from bowing. To provide a slight overhand, set the 2×8 riser boards at an angle of about 15 degrees. To allow troweling of the entire surface of the tread, the bottom edges of the boards are beveled. The key to building forms for steps or platforms is to keep the reverse of the forms in mind when creating an angle or beveled edge. Visualize how the project will look when the forms are pulled, not as they look before the pour is made. This will save time, money, and much effort.

WATERPROOFING

Waterproofing the outside of foundation walls, especially basement walls, is required in most geographic locations. The principle of waterproofing is to position a waterproof membrane around the entire foundation so that water cannot flow or trickle into the interior of the structure. Waterproofing is applied to basement walls that are below grade.

Before waterproofing is applied, drain tile is placed around the exterior of the foundation at the point where the footing and wall join. The drain tile is placed at a slope of about 1 inch in 20 feet, and spaced ¼ inch apart. Strips of tar paper cover the joints. Plastic drain tile can be used. This tile comes in 100-foot rolls and has openings located throughout it so water can flow in and then be carried to an outlet where it is drained off. These tile materials are covered with 6 to 8 inches of coarse gravel or crushed stone. Once this operation has been completed, the waterproofing is applied.

For block foundations, a coat of cement plaster is applied, followed by several coats of an asphalt material. For poured foundations, use asphalt or hot bituminous material for waterproofing. Polyethylene sheeting is used frequently. This material holds water away from the foundation if it is installed correctly with the joints lapped at least 6 inches and glued together with a special polyethylene adhesive. Care must be taken not to puncture or tear the sheeting when it is being installed or when the foundation is being backfilled.

WOOD FOUNDATIONS

Wood foundations are used in some areas of the country, such as Alaska, due to the permafrost and the inability to dig into the soil to any depth. Wood foundations are also less costly and new methods of weatherproofing wood have increased the wood's resistance to insects and fungus growth.

Wood foundations have been approved by the Federal Housing Authority (FHA), the Department of Housing and Urban Development (HUD), and the Farmers Home Administration (FHA) as durable and long-lasting. All wood used for foundations should be rated with the mark that reads "AWPB-FDN." This mark assures that the lumber meets the requirements of code organizations and federal regulatory agencies.

Some wood foundations are prefabricated and are sent to the job site ready to be set in place. If the foundation is not prefabricated it can be constructed on the site from 2-inch material and weather treated plywood. Use nails made of silicon, bronze, copper, or hot-dipped zinc-coated steel. Seal the joints between the panels with a special caulking compound. If a basement floor is to be poured, a porous gravel base must be laid down and covered with polyethylene film (6 mil thick); a screed board must be attached to the foundation wall.

A double top plate, made up of two 2 × 4s nailed together and placed on top of the side wall framing, must be used for the first floor frame. The frame should be attached in such a way as to transfer the inward forces to the floor structure. Before backfilling around the wooden foundation, a 6 mil polyethylene moisture barrier must be installed on all walls that are below grade. Make sure the joints in the film have at least a 6-inch overlap. Seal the joints with special polyethylene adhesive.

Specifications for wooden foundations and the standards for their construction should be obtained before attempting construction. The National Forest Products Association publishes a manual on how to build wooden foundations, including codes for the lumber to be used.

SPECIAL PROBLEMS WITH FOOTINGS AND FOUNDATIONS

As with any construction phase, a number of problems can arise. Many can be eliminated by taking the time to accurately locate and lay out the building site and by checking to make sure that the foundation is square and capable of bearing the weight of the structure.

Some construction problems, such as those relating to climate, cannot be controlled but must be resolved. Cold-weather construction requires special steps to prevent additional problems later on. If concrete is poured on days when the temperature falls below 40 degrees, a shelter must be erected over the pour site, or a special mixture must be used for the concrete.

On big pours, the entire building area may have to be enclosed in a plastic structure and heated. This can be done with scaffolding, lumber, and plastic sheets. If a shelter is not built, the freshly poured concrete must be covered with tarpaulins or plastic to retain the heat from the hydration process and to slowly bring about the curing process without stress or damage to the concrete. In extreme cases in which the pour must continue, the water can be heated up to 180 degrees. If the water is heated beyond this point, it may cause the concrete to set instantly. The sand that is being used for these cold pours should be heated to melt any frozen lumps that might have been created. Admixtures, such as antifreeze and accelerators, can be added to the water to slow down or speed up the curing process. A rule for cold-weather construction involving concrete is to delay the pour if you can, and if you can't, take all precautions to ensure that the pour will not have to be repeated.

INSULATION FOR SOUND AND WEATHER

Insulation is installed in structures to control the transfer of heat or sound. The selection of the appropriate insulation material depends on the geographic location of the structure, the cost of materials, the level of insulation desired, and the kind of structure to which it is attached.

Heat transfer is the controlling factor when insulating against either heat loss in the winter or heat invasion in the summer. Heat transfer occurs by air infiltration and by transmission through the structure. Air infiltration is leakage through cracks and other open spaces. The volume of air entering the building is offset by the air leaving it. Since the temperatures of the two air volumes are normally different, there is a transfer of heat in any air exchange. The exchanged air requires energy to either heat or cool it, depending on the desired effect.

Air transmission occurs through walls, windows, doors, and roofs. Heat is transmitted by molecular conduction through wall materials, by convection of air currents that circulate in open spaces within walls and roof construction, and by radiation in which the energy of the sun is transmitted from one surface to another.

PRELIMINARY STEPS

The first step in insulating is determining the areas that are to be insulated. The logical areas are walls, ceilings, and floors. It is best to keep the insulation as close to the heating or cooling source as possible. For example, when insulating an attic, the insulation should be placed in the attic floor rather than in the roof structure. This will keep the hot or cool air in without using energy to maintain the unused space in the attic. For a slanted or vaulted ceiling, insulation should be placed between the rafters, but adequate space must be maintained between the insulation and the sheathing for air circulation. Insulation should be placed in floor joists, where the structure projects over nonheated areas such as porches, garages, or crawl spaces. (See Figure 4–49.) Basements used

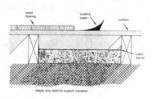

(Figure 4-49)   Insulation

as recreation or living areas should be insulated. The purose of insulating interior walls is not to save energy, but to reduce sound penetration.

If the structure does not have a basement, the floors over the unheated space must be insulated at the same level as the side walls. When insulating a crawl space, apply a vapor barrier. If the structure does not have a crawl space, as is the case with many older homes, rigid insulation can be applied around the outside of the foundation.

The second step in insulating is selecting suitable insulation material. The selection of the material is based on the location of the area to be insulated, the amount of space that is available for insulation, and the R-value (which represents the material's ability to resist heat transfer) that is desired. For example, if you want to insulate a small crawl space, and you want a high R-value, the best type of insulation to use in this case would be rigid insulation panels made from foamed plastic.

SOUND INSULATION

Sound insulation is installed to control the level of noise that penetrates an area. Sounds in homes come from voices, televisions, radios, musical instruments, power equipment, service equipment (such as air conditioners), and external sources (such as traffic). The key to controlling these sounds is either to eliminate the source, which in most cases is not an option, or to control the transmission of the sounds.

Sound transmission is based on sound waves striking a surface, such as a wall, floor, or ceiling. The sound waves cause the walls to vibrate, and the vibration transmits the sound to an adjoining area.

The National Bureau of Standards has established a sound classification system that rates construction materials and methods. A high sound transmission class (STC) rating is desired for walls that must control noise. Certain areas are logical choices for noise control. Bathroom walls, for example, should have a high STC rating, as should walls that divide bedrooms from living areas.

There are three ways to create walls with high STC ratings. The first is to select sound-reducing material, such as structural insulation board, which is made from wood and cane fibers. This board comes in ½ inch thickness and 4 × 8 foot sheets. It is made expressly for sound control. A second alternative is to construct double walls. Double walls are frequently found in apartment construction and are insulated in the same manner as for thermal insulation. The third alternative is to soundproof the floors and ceilings. Floors can be soundproofed by adding a layer of 4-inch blanket insulation between the floor joists. The insulation is held in place by stapling it to the joists or by screwing metal channels across them. The addition of floor covering further reduces noise transmission.

Ceilings can be soundproofed by using patented metal clips to attach the ceiling material to the joists. Another form of noise control for ceilings is the use of accoustical tile or suspended ceilings.

Sound is also transmitted through metal duct systems (which carry warm and cold air throughout the house). To reduce the transmission of noise, exposed ducts that are accessible, such as those in the basement, can be wrapped with insulation that is held in place by a wooden framework.

THERMAL INSULATION

Thermal insulation is rated by using a factor called the *R*-value. The *R*-value represents the material's resistance to heat transfer, as opposed to its ability to conduct heat. Good insulation material will have a high *R*-value. The American Society of Heating, Refrigerating and Air Conditioning publishes data that outlines the *R*-value for commonly used insulation and building materials.

Thermal insulation generally falls into four general categories: rigid boards, flexible batts or blankets, loose fill, and reflective material. The first three work on the principle of trapping air in thousands of tiny pockets within the material to prevent heat passage in either direction. The reflective type has foil on the surface that reflects the heat. Some insulation types incorporate both the entrapment principle and the reflective principle through the use of reflective, foil-covered batts. Many local building codes now call for a minimum insulating value for walls, ceilings, windows, and basements. In addition, some builders are using super-insulation or double-wall construction. Super insulation consists of an extra thick wall, rather than the standard 2 × 4 stud construction. The walls are 2 × 6 or 2 × 8 lumber with insulation of the appropriate thickness and in the form of either rigid foam or batts. The outside sheathing is rigid foam instead of standard blackjack, which is a rigid insulation that's covered on one side with a black coating of asphalt. A polyethylene vapor barrier is installed over the insulation for maximum seal.

Double-wall construction consists of 2 × 4 studs set in the standard manner for framing, except that they are set on 24-inch centers. A second wall made of 2 × 4 studs is set inside the first wall. These studs are on 24-inch centers but are staggered so that they don't line up with the outside studs. This prevents conduction of heat or cold air. A dead space of approximately 2 inches is maintained between the two walls. Each wall is insulated, and a vapor barrier is installed on the inside of the interior wall. Cost is an obvious factor in construction of this type; planning is another, since the foundation, utilities, and exterior openings are affected.

FILL INSULATION

Fill insulation consists of granulated rock wool in the forms of nodules or pellets, granulated cork, expanded mica, and other material. This insulation is blown through large tubes or pumped into open spaces in the walls and ceilings, such as the stud spaces in frame walls and ceiling joist spaces. Fibrous mineral wool and glass wool are also available in loose form so that they can be packed by hand into open spaces. This form of insulation is frequently used in older homes, since it can be installed with minimum work.

FLEXIBLE INSULATION

Flexible insulation comes in quilts or blankets and batts. The quilts or blankets consist of a fibrous material such as treated wood fiber, hair, felt, flax fiber, or shredded paper stitched between sheets of waterproof paper to form a flexible material available in various thicknesses.

Another form of flexible insulation is made from mineral or rock wool, a fluffy, noncombustible product weighing about 6 pounds per cubic foot. It is usually furnished in batts 15 × 24 or 15 × 48 inches to fit between studs and ceiling joists spaced 16 inches center to center. The thickness is ordinarily about 4 inches to fill the space completely between 2 × 4 studs, but 6-inch batts are available for use with 2 × 6 studs. For basement insulation, the batts come 2 inches thick. Batts also come 12 inches thick for use in attics. The batts are furnished without any backing or with waterproof paper cemented to the back and projecting about 2 inches on either side to provide nailing flanges. This paper serves as a seal and vapor barrier. To be effective, the end joints must be tight with no gaps between the end of the insulation and the area being insulated.

REFLECTIVE INSULATION

Reflective insulation is placed in air spaces and functions by reflecting a large percentage of any radiant energy that strikes it. The materials used for reflective insulation are very thin tin plate, copper, or aluminum sheets, or aluminum foil on the surface of rigid fiberboards or gypsum boards. A common form of reflective insulation is aluminum foil mounted on asphalt-impregnated kraft paper, the strength of which may be increased by the use of jute netting. This material is used for lining air spaces or for curtains to increase the number of air spaces in a given overall space. To be effective, the edges of such curtains must be tightly sealed. The material comes in 4 × 8-foot sheets or rolls of batts of varying widths.

RIGID INSULATION

Rigid insulation is made of fiberboards. The fibers are lightly compressed and are covered on one side with a reflective aluminum foil coating or a black coating of asphalt. This insulation is commonly called "blackjack" and is installed over the studs on the outside walls of the structure as sheathing. It usually comes in 4 × 8- or 4 × 9-foot sheets and is ¾ inch thick.

VAPOR BARRIERS

Vapor barriers are impervious to moisture. Commonly used materials are plastic film, metal foil, or asphalt between layers of brown paper. The most commonly used of the three is the plastic film. The vapor barrier is installed on interior surfaces of the interior walls and ceilings.

VENTILATION

A vapor barrier is installed to prevent moisture from getting into the structure through condensation. This precaution is one that will save money and prolong the life of the structure, but one more factor—ventilation—must be accounted for to complete the insulation process. A building must be able to breathe, especially if it has an unheated attic or a low-pitched or flat roof. A building can be made to breathe by installing ventilators on the ends and top of the roof. On gabled roofs, the ventilators are located on the gable ends. On hip roofs, the ventilators are installed right into the roof. If installed properly, there is no problem with leakage. The eaves of buildings should have soffit vents installed along with the roof ventilators. These two vents will allow air to circulate through the attic and vent off the attic air. (See Figure 4–50.) This air needs to be circulated whether it is winter or summer. A special concern is that if adequate venting is not provided, the warm air can escape from

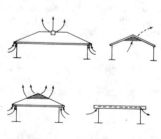

(Figure 4-50) Attic vents

the rooms below, move up to the bottom of the roof, melt the snow and ice, which in turn runs down the shingles to the gutter where it freezes again forming an ice dam. This can cause the water to back up under the shingles and leak into the building. Also, be sure that insulation does not block the airway from the soffit vent into the attic. Special baffles are available that attach to the rafters to ensure that adequate air will pass through.

If a wall has inadequate cold or outside ventilation, it can be corrected by installing patented ventilators. These units are simply metal tubes through which air passes. The outside of the tubes have tiny louvers that

(Figure 4-51)   Patent vents

direct the flow of air. They are installed by drilling holes through the siding and sheathing and then pressing the tubes into place. (See Figure 4–51.) For maximum ventilation, install a ventilator at the top and bottom of each stud space.

ROOFS

Roofs are composed of three major components: the skeletal framing or supporting members, a stiff sheeting to support the outer skin, and a waterproof outer layer of roofing material. The framing may be rafters or trusses. The supporting decking may be a concrete slab, metal roof decking, poured or precast gypsum units, wood sheathing or fiberboard. The roofing may be any number of materials.

Careful planning is required to ensure that each of these building covering components is adequate to support and protect the structure upon which it is located. Each of these three components is described in depth to assist in the planning and selection of a well-designed, economical, and pleasing roof system.

GUTTERS

Gutters control water runoff so it does not cause damage to the roof, structure, foundation, or surrounding area. Gutter systems come in many different sizes and materials. The size of a gutter is based on the square footage of roof that is being drained into the gutter. As a general rule, a roof area of up to 750 square feet can be handled by a 4-inch-wide trough. For areas between 750 and 1,400 square feet, a 5-inch trough should be used; for larger areas, a 6-inch trough should be used.

Downspouts, or conductor pipes, are also sized based upon the roof area. For roofs up to 1,000 square feet, downspouts with 3-inch diameters are sufficient. Larger areas require 4-inch downspouts.

Gutters are made from wood, galvanized iron, aluminum, or vinyl plastic. The most commonly used is galvanized iron, that may or may not be finish painted. Continuous gutters that are made to fit specific jobs are gaining in usage. This type of gutter comes in a variety of colors and requires a special machine to form the sheet metal into the gutter shape. The advantage of this system is that there are no joints to complete or seal, only corners. A standard gutter system has gutter troughs, inside and outside mitered corners, joint connecters, pipes, brackets, and other items, such as outlet tubes. To install a system like this, the gutter trough is cut to the proper length and the supporting hardware is placed on the trough so that the whole system can be hung in place. The parts of the gutter system are held in place with sheet metal screws or pop rivets. All joints are sealed with special gutter seal that comes in either a tube or a cartridge.

Gutters are hung with a slope so that the water will move quickly through the system. The gutters should slope 1 inch for every 12 to 16 feet of length.

HIPS AND RIDGES

Hips and ridges must be tight and well constructed in order to avoid leakage. This is true whether asphalt or wood shingles are used, though the methods used to create the special shingles are slightly different.

Asphalt hip and ridge shingles can be purchased from roofing suppliers. They can be made very easily by cutting 9 × 12-inch pieces from either square-butt shingle strips or from rolled roofing that matches the color of the shingles. After the shingles have been cut, they should be bent lengthwise in the centerline. Begin installation at the bottom of the hips or at one end of the ridge.

Lap the units to provide a 5-inch exposure. Drive one nail on each side, 5-½ inches back from the exposed end and 1 inch from the edge.

When laying wood shingles, the hip and ridge shingles are harder to create and install than the asphalt shingles, but with care this process can be completed with waterproof results. Wood hip and ridge shingles can be purchased from roofing manufacturers, or they can be made by selecting regular shingles of approximately the same width as the roof exposure. Lay out two lines on either side of the hip or ridge where the shingles are to be laid. Place one shingle over the other to create a cap effect. These shingles are then nailed into place using extra-long nails to achieve adequate penetration.

MATERIALS

ASPHALT ROOFING PRODUCTS

Asphalt roofing products are commonly used in construction due to their availability and cost. These roofing products include three broad categories: saturated felts, roll roofing, and shingles.

Saturated felts are used under shingles for sheathing paper and for laminations in constructing a built-up roof. They are made of dry felt soaked with asphalt or coal tar. The saturated felts come in various weights from 10 to 50 pounds. The weight indicates the amount of material necessary to cover 100 square feet of roof surface with a single layer. The roofing felts are sold by rolls that normally are 36 inches wide and 50 or 100 feet long.

Roll roofing is made of an organic or inorganic felt saturated with an asphalt coating and an additional viscous bituminous coating. Finely ground talc or mica may be applied to both sides of the felt to produce smooth roofing. Mineral granules in a variety of colors are rolled into the upper surface while the surface is still soft. The granules protect the underlying surface from the sun and increase the material's fire resistance. Roll roofing comes in weights of 75 to 90 pounds per 100 square feet (expressed as lb/square) and with one surface completely covered with granules or with a 2-inch plain-surface selvage along one side to allow for laps.

Shingles, sometimes called composition shingles, are available in several patterns in strip form or as individual shingles. The shingles are manufactured on a base of organic felt or a fiberglass mat. The felt or mat is covered with a mineral stabilized coating of asphalt on the top and bottom. The top side is coated with mineral granules of specified color. The bottom side is covered with sand, talc, or mica.

The most common shape of asphalt or fiberglass shingles is a 12 by 36 inch strip with the exposed surface cut or scored to resemble three 9 × 12-inch shingles. These are called strip shingles. A lap of 2 to 3 inches is usually provided over the upper edge of the shingle in the course directly below. This is called the head lap. Strip shingles are produced in two designs to create different patterns. These two designs are straight-tab or random-tab. The random-tab is designed to give the illusion of individual units or wood shakes.

Most strip shingles have a strip of adhesive spaced along the concealed portion of the strip. This adhesive helps the shingles to become self-sealing once the sun has warmed the adhesive to the point where the shingles stick together.

DRIP EDGE

Drip edge is a metal strip that goes along the edge of the eaves and rake. Drip edge is made from 26 gauge galvanized steel and comes in a variety of shapes. The purpose of drip edge is to create an edge that bends downward from the roof edge from which water can drip freely without touching the cornice construction. At the eaves, the underlayment is laid over the drip edge; at the rake, the underlayment is laid under the drip edge. One side of the drip edge is flat and approximately 3 inches wide. This side of the edge goes toward the structure, and the roofing is laid over it. This width prevents the

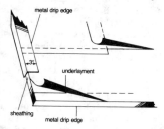

(Figure 4-52)   Drip edge

water from backing up and running under the shingles where it can leak into the structure. (See Figure 4–52.)

FLASHING

Flashing is installed for the purpose of creating a waterproof joint. It is placed where there is a possibility of water leaking in around an area where the roofing material is disrupted in its layout. This occurs most frequently around masonry walls, chimneys, and valleys on roofs. (See Figure 4–53).

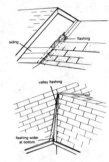

(Figure 4-53)   Flashing

Flashing is made of heavy flexible plastic sheets, roofing material, or strips of sheet metal such as galvanized steel, aluminum, copper, zinc, various alloys, or clad metals. Flashing comes in different shapes or in a flat metal roll that can be worked into the shape of the area in which it will be used.

Flashing for use on eaves is a strip of smooth or mineral-surfaced roll roofing that is cut to the width that extends from the roof's edge to about 12 inches inside the wall line. The strip is installed over the underlayment and metal drip edge. This flashing prevents leaks from water backed up by ice dams on the roof.

Chimney Flashing. Special precautions must be taken to make the roof leakproof around the chimney opening while allowing for movement of the structure. A method for solving this problem is to secure the base flashing to the roof deck and fasten the cap flashing to the chimney. The angle between the higher portion of a chimney (or other projection) to a sloping roof is protected by a saddle, or cricket. A saddle is usually formed out of ¾ inch exterior plywood; if the saddle is small, it doesn't require framing. For larger saddles, framing is required, using standard roof framing techniques. The saddle is then covered with corrosion-resistant sheet metal or mineral-surfaced roll roofing. The valleys that are formed around the saddle are sealed using the same method used for regular roof valleys.

When applying the base flashing, lay the shingles up to the front face of the chimney; then lay out and cut (from 90 pound mineral-surfaced roofing) the front section of the flashing. If a saddle is not used, a section of flashing can be cut out for the back side of the chimney. The side sections of the flashing are then cut and put in place. All joints are sealed with asphalt plastic cement. Many roofers prefer to use metal flashing for the base. It is installed in the same manner as the mineral flashing, making sure all joints are securely sealed.

Cap flashing is made of sheet metal strips that set into mortar joints of the chimney. They are installed as the chimney is being built. When completing the flashing job, bend the metal cap flashing strips down over the base flashing. The cap flashing on the front of the chimney should be one continuous piece, while the flashing on the sides is stepped up the roof in sections and bent to conform to the slope of the roof.

Flashing Shingles. Flashing shingles are used where a roof joins a vertical wall. Flashing shingles are 26 gauge galvanized metal pieces that are cut to size. They should be at least 10 inches long and 2 inches wider than the exposed face of the regular shingles. When installing the metal shingles, the 10-inch length is bent so that it will extend 5 inches over the roof and 5 inches up the wall.

Basic Procedures for Installing Flashing Shingles

1. *As each course of shingles is laid, a metal flashing shingle is installed and nailed at the top edge of the roof shingle that abutts on the vertical wall.*

2. *Do not nail the metal flashing to the wall as settling of the roof frame can cause damage to the seal.*

3. *Wall siding is installed after the roof is completed and serves as cap flashing. Position the siding just above the roof surface, allowing enough clearance to paint the lower edge.*

4. *Where the asphalt shingle rests on the metal flashing, embed the end of the shingle in asphalt plastic cement to make sure that the shingle stays in place.*

Valley Flashing. Valley flashing is used where roofing materials come together when the roof is sloping. Water drainage is heavy at this point, and leaks are easily created unless maximum precautions are taken. Valley flashing may be metal or rolled roofing material.

Basic Procedures for Installing Valley Flashing

1. *When metal flashing is used for valleys, it is first cut to length.*

2. *Set the flashing in place and form to fit the valley.*

3. *Nail in place using standard roofing or galvanized 6d nails.*

4. *Flashing using roofing materials requires that a strip be cut 18 inches wide. This strip is centered in the valley and laid with the mineral surface down.*

5. *A second strip is cut 36 inches wide and laid down the center of the valley on top of the 18 inch strip, with the mineral side up.*

6. *Any joints are lapped at least 12 inches and are sealed with plastic asphalt cement.*

7. *When laying the strips, nail one side, press the material firmly into the valley, and then nail the other side.*

8. *Before installing the shingles, snap a chalk line down the center of the valley and on each side of the valley. The outside lines mark the width of the waterway. The valley should be 6 inches wide at the ridge and become gradually wider. The lines move away from the valley at the rate of ⅛ inch for every foot as they approach the eave. The outside lines are then used as guides in trimming the last units of the shingles as they fit into the valley.*

9. *Another method of installing flashing in a valley is to place a 36-inch strip of roll roofing (50 pounds or heavier) down the center of the valley and then lay the shingles right into the valley. Strip shingles are the only type that will work with this method. The shingles are laid into the valley and then woven together with the excess trimmed off the ends. Another finish method is to lay a closed cut valley. The closed cut valley has the shingles on the right stopping 2 inches from the center of the valley. The shingles intersect with cut shingles fit into the valley, and the extra tab is slipped behind the shorter shingles. This method creates a tight bond. Also, when cutting the shorter shingles in, they should be embedded in plastic asphalt cement.*

10. *This same method of installing flashing in a valley can be used when laying shingles between a main roof and a gable dormer.*

ROLL ROOFING

This roofing material comes in rolls 36 inches wide—a 17-inch wide section is covered with granular material and a 19-inch wide section (the selvage) is smooth. Roll roofing comes in a variety of weights, surfaces,

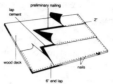

(Figure 4-54)   Roll roofing

and colors. The most common system used to install roll roofing is the double-coverage method. (See Figure 4–54.)

Basic Procedures for Installing Roll Roofing

1. *Double-coverage roll roofing is normally laid parallel to the eaves.*

2. *The starter strip is made by cutting off the granular-surfaced portion of the roll roofing that has been cut to length for the roof.*

3. *The starter strip is nailed along the top and bottom. The lower row of nails should be driven 1 inch above the lower edge of the strip. The upper row of nails is located 4¾ inches below the upper edge.*

4. *Cover the starter strip with asphalt cement and overlay a full width strip. Nail the strip in place by driving one row of nails 4¾ inches from the upper edge and a second row 8½ inches below the first row. The nails should be spaced about 12 inches apart.*

5. *Each successive strip should be installed so that it overlaps the full 19-inch selvage area. The sheet should be nailed in place. Then the top should be turned back so that cement can be applied to it. The cement should be applied to within about ¼ inch of the granular surface. The overlaying sheet must be pressed firmly into the cement using a roller or some other pressure device.*

6. *Flashing and hip and ridge caps are installed where needed, using the procedures outlined in the sections discussing these items.*

SHINGLES

Shingles are made from asphalt, wood, slate, fiberglass, and metal. The most commonly used materials are wood and asphalt. Asphalt shingles are sometimes called composition shingles, and they come in several patterns in strip form or in individual shingles.

As mentioned earlier, strip shingles commonly come in 12 × 36-inch strips with the exposed surface cut or scored to resemble three 9 × 12-inch shingles. Most strip shingles have a factory-applied adhesive spaced at intervals along the concealed portion of the strip. These strips of adhesive are activated by the warmth of the sun and hold the shingles firm through winds, rain, and snow.

Single shingles are made from asphalt or fiberglass and lock together using several different patterns. The layout for single shingles is the same as for strip shingles. Single shingles are used primarily to create a particular pattern or to provide added strength against high winds.

Basic Procedures for Installing Strip Shingles

1. *Estimate the amount of material required. Shingles are sold by the square, which provides 100 square feet of coverage.*

2. *For runs of less than 30 feet the shingles can be started at either end of the structure. If the roof is longer than 30 feet, it is best to start from the center.*

3. *To keep the shingles accurately aligned, snap a number of chalk lines between the eaves and the ridge. These lines are used as reference points to show where the next course is to be laid. The position of these lines depends on the type and size of shingles used. Experienced workers lay out guidelines every fifth or sixth course. Inexperienced workers should set guidelines every two or three courses.*

4. *Strip shingles are nailed using 12 gauge galvanized steel nails with barbed shanks or aluminum nails. The nails should be long enough to penetrate the shingles and the thickness of the sheathing. A normal length is 1¼ inches for new roofs and 1¾ inches for reroofing. The placement of the nails on the shingles is very important. Manufacturers generally supply application diagrams with the shingles. For strip shingles, normally four nails are driven along the top of the shingle; if a 5-inch exposure is used, then the nails are placed ⅝ inch above the tab slits and 1 inch from each end. Always position the shingle and start nailing from one side to the other. If you nail one end and then the other, you can create a buckle in the middle of the shingle. Drive the nails in straight so that they will not cut*

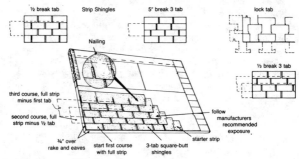

(Figure 4-55)   Installing shingles

the surface of the shingles. Nails should be driven flush with the shingles, not below the surface.

5. *Pneumatic powered staplers are frequently used for roofing. The staples are normally 16 gauge and ¾ inch long with extra-wide crowns. When using a power stapler, set the pressure properly to ensure that the staples will be set flush with the shingles. Staples are normally placed ⅜ inch below the adhesive strip.*

6. *A starter strip is applied to back up the first course of shingles and fill in space between the tabs. (See Figure 4–55.) The starter strip is a 9-inch strip of mineral-surfaced roofing cut from rolled roofing that matches the color and surface texture of the strip shingles. The starter strip is positioned so that it overhangs the drip edge slightly. It is nailed in place by locating nails 3 to 4 inches above the edge. The nails should be spaced so they will not be exposed by the slits in the tabs of the first course of shingles. Some roofers like to use an inverted row of strip shingles for the starter. If self-sealing shingles are used, the tabs may be cut off and the shingles set right side up. The adhesive serves to hold the first course of tabs in place.*

7. *The first course is started with full shingles. Succeeding courses are then placed with either full or cut strips, depending on the type of shingle or the laying pattern. Three-tab square-butt shingle strips are commonly laid so that the cutouts are centered over the tabs in the course directly below; thus the cutouts in every other course will be exactly aligned.*

8. *To create the centered pattern, the second course starts with a strip that has had 6 inches cut from it. The third course is started with a strip with a full tab removed, and the fourth with half a strip. The fifth course starts with a strip that has had 24 inches removed. Course six starts with 30 inches removed, and the seventh course starts with a full shingle. The layout pattern is then repeated until the roofing has been completed.*

9. *Normally, the courses are laid out using partial shingles until a full shingle is needed. This layout pattern usually involves six courses. This number of courses allows you to keep the shingles in line as well as keep the courses within easy reach for nailing.*

10. *After reaching the ridge, go to the other side of the roof and repeat the entire process. Trim any excess material off of the last course of shingles laid. You are now ready to lay the ridge shingles. (For cutouts and flashing, see the sections on flashing and hips and ridges.)*

Wood Shingles and Shakes. Wood shingles are made from cypress, redwood, or cedar. Most wood shingles used in Canada and the United States are made from western red cedar, which has natural oils that resist moisture and produce a pleasant color, ranging from yellow to brown. Grading for wood shingles is based on the amount of knots processed, the grain pattern, and whether the shingles have been cut from heartwood. Shingles graded No. 1 are cut from heartwood and have vertical grain with few knots. Economy shingles contain increasing amounts of sapwood, twisted grain, and knots. Wood shingles come in lengths of 16, 18, and 24 inches and in widths from 3 to 14 inches. Four bundles of wood shingles are needed to cover one square (100 square feet). The exposure of the shingles depends on the slope of the roof and requirements of local building codes. A rule of thumb for laying wood shingles is that at least four layers or courses of shingles should be used over the entire roof area.

Wood shakes are similar to shingles except that they are split, rather than sawed, from 100 percent heartwood bolts. There is only one grade of red cedar shakes. Individual shakes are split from the bolt with a heavy steel-bladed tool called a froe. They are produced in lengths of 18, 24, and 32 inches. Coverage for each size varies depending on width and lay of the shake. Texture and color also vary from one shake to another. Shakes may be straight split, tapersplit, or hand split and resawed. Straight-split shakes are split from the bolt to the desired thickness and are not tapered. For tapersplit shakes, the blocks are turned end for end after each split to produce the taper. Hand-split and resawed shakes are split into the desired thickness and then passed at an angle through a band saw. This produces a tapered shake that has a split face and a smooth-sawed back.

Basic Procedures for Installing Wood Shingles and Shakes

1. *Estimate the amount of shingles or shakes needed.*

2. *Use rust-resistant nails to fasten wood shingles and shakes. Hot-dipped, zinc-coated, steel nails are the best for this work. The nails must be of sufficient length to ensure adequate holding power and are best driven using a shingler's hatchet. The hatchet has a gauge for laying out the shingles and can be used for splitting and trimming.*

3. *Shingles and shakes may be applied over solid or spaced sheathing, with 1 × 3, 1 × 4, or 1 × 5 fir or pine strips laid horizontally at the same distance on center as the shingle or shake exposure. Solid sheathing and felt make the roof more airtight and watertight; however, water that penetrates to the underside of the shingles cannot evaporate and will cause cupping and rotting. If a solid sheathing is desired under wood shingles, lay 1 × 3 strips over the sheathing to allow air to circulate under the shingles or shakes.*

4. *Each shingle or shake is nailed to the stripping or sheathing using only two nails to avoid splitting. The nails should be driven 1 to 1½ inches above the butt line, so that the next course covers the nails, and not more than ¾ inch from the edge of the shingle or shake.*

5. *The first course of shingles or shakes at the eaves should be doubled or tripled. All shingles or shakes laid on the roof should be spaced ¼ inch apart to allow for expansion when they become rain soaked.*

6. *The second course of shingles or shakes should be nailed over the first layer so that the joints in each course are at least 1½ inches apart.*

7. *Care should be taken in successive courses to break the joints in a variety of ways so that they do not match up in three successive courses.*

8. *To keep the courses straight, nail a board temporarily in place to hold the shingles or shakes in position until they have been nailed.*

9. *A chalk line should be used to check alignment every five or six courses.*

10. *Check that the shingles or shakes are laid parallel to the ridge.*

UNDERLAYMENT

Underlayment is a thin cover of asphalt-saturated felt or other material. It protects the sheathing from moisture until the shingles are laid, provides an extra layer of weather protection by preventing the entrance of wind-driven rain and snow, and prevents direct contact between shingles and resinous areas in the sheathing.

Underlayment should not be coated or made of heavy felt that might act as a vapor barrier, since this could trap moisture between the sheathing and the roofing material. A 15-pound underlayment is most commonly used for roofing.

Underlayment is installed by using a 2-inch lap at all horizontal joints and a 4-inch side lap at all end joints. The process of installing underlayment is called "blacking out" a roof. The underlayment is held in place with thin metal disks that are nailed with roofing nails to the sheathing. When valleys are blacked out, the lap joints should be 6 inches on each side of the centerline. This 6-inch lap also applies to hips and ridges.

OTHER ROOFING MATERIALS

There are a number of other roofing materials and forms that are used in construction, though they are not used as often as those just discussed. Among these materials are tile roofing, mineral-fiber shingles, galvanized sheet metal roofing, aluminum roofing, terne metal roofing, aluminum shakes, built-up roofs, slate shingles, and cement tile. All of these materials have unique characteristics that warrant their use at different times and in specific locations. The selection of these materials depends a great deal on the architectural design and cost of the project. In some cases, if cost is not a factor and the design is somewhat different from that normally used in the surrounding geographic region, a combination of roofing materials may be installed.

PURLINS

Purlins are ledgers that support long runs of rafters and are required when the distance spanned by the rafters exceeds the maximum allowed by code. They are usually made of 2 × 4 stock and rest on plates located over supporting partitions. Bracing under the purlins may be placed at any angle to transfer loads from the midpoints of the rafters to the support below. (See Figure 4–56.)

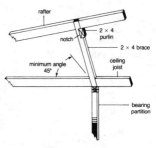

(Figure 4-56)   Purlin

RAFTERS

Rafters are the frame members upon which the sheathing and roof material rest. They must be accurately cut and fit into place to minimize movement of the roof caused by structural settling or wind. Such movement could cause the roofing material to leak.

If the roof is complicated the layout for rafters will generally be included in the building plans. (See Figure 4–57.) If the roof layout is simple, it is often left up to the carpenter to visualize the required

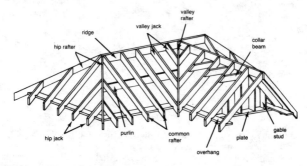

(Figure 4-57)   Rafter plan

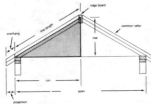

Common Rafter Layout

(Figure 4-58)   Common rafter layout

roof framing. This is not difficult to do, since the wall framing is already in position and only the slope of the roof and the amount of overhang need to be determined to complete the layout. (See Figure 4–58.) If a framing plan is not provided, it is best to prepare a simple plan by placing a piece of tracing paper over the scaled house plans. The drawing does not have to be elaborate—a single line can represent the rafters. The spacing between the rafters should be drawn accurately, as should the hips and valleys (at a 45 degree angle); the jack rafters must be parallel to the common rafters.

Pitch, rise, and run must be determined when laying out a roof. The pitch, rise, and run can be determined from a basic formula when two of the quantities are known. The formula is $R = P \times S$, where R represents the rise; P, the pitch; and S, the span. For example, the pitch for a building that is 24 feet wide and has a rise of 8 feet can be calculated as follows: $8 = P \times 24$; $P = \frac{8}{24}$, or $\frac{1}{3}$.

Using the same 24-foot span, the rise in inches for each foot of run can be determined as follows. The rise is 8 feet (or 96 inches). The run is one-half the span (in this case 12 feet). Therefore, the rise per foot of run equals 96 inches divided by 12, or 8 inches.

The rise in inches per foot of run is always the same for ordinary pitches:

Pitch	1/2	1/3	1/4	1/6
Rise per foot of run (in inches)	12	8	6	4

COMMON RAFTERS

Common rafters are straight-cut and, as their name implies, they are the most commonly used rafters. To lay out a common rafter, a practical application of geometry is required.

Generally, one of two methods is used for laying out rafters.

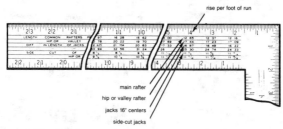

(Figure 4-59) Framing square rafter tables

Both methods require the use of the framing square, and both methods are based on the relationship of the sides of a right triangle. The horizontal distance represents the base of the triangle, the vertical distance represents the altitude (or height), and the length of the rafter represents the hypotenuse. The geometric formula for finding the third side of a triangle when two sides are known is not used, since this process is time-consuming and complicated. The first practical method is to use the table on the framing square; the other method is the direct-layout method.

The tables on the square are based on the rise per foot of run (length). Numbers in the tables show the length of common rafters for any rise. (See Figure 4–59.) The length of the common rafter is the shortest distance between the centerline of the ridge and the outer edge of the plate. This length is taken along the measuring line. Inches and sixteenths of an inch are found on the outside face of the square on both the blade and the tongue. The first line on the blade of the square gives the lengths of common, or main, rafters per foot of run. The seventeen main rafter tables begin under the 2-inch mark and continue through 18 inches. To find the length of a common rafter, multiply the number of feet of run by the length given in the table. As an example, if the rise equals 8 inches per foot of run (⅓ pitch) for a building 20 feet wide (span), the run is 10 feet. Locate the 8-inch mark on the body of the square. On the first line under the 8 is the number 14.42. This is the length of the common rafter in inches per foot of run. Multiply 14.42 by 10 feet of run to arrive at 144.2 inches of rafter. Divide 144.2

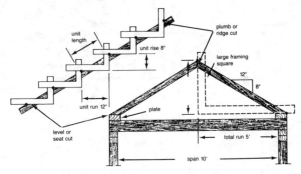

(Figure 4-60)   The step-off method

inches by 12 inches per foot; the common rafter length is 12.01, or 12 feet for practical use.

The step-off method is another way of laying out common rafters. (See Figure 4–60.) When using the step-off method, the framing square is held with the unit rise along the edge of the rafter on the tongue of the square, and the unit run on the body. A mark is made on each side of the square, and the square is stepped off in the usual manner, but instead of marking at the edge of the rafter for a full unit of run, a mark is placed at the inch mark equal to the fraction of the run. While the step-off method of laying out a rafter is easy and theoretically correct, it is not as accurate as the unit method, because a small amount can be gained or lost at each step. A gain of only ¹⁄₁₆ inch on each of twelve steps will result in a rafter ¾ inch too long. To avoid gaining length on each step, a sharp pencil or knife should be used to mark each step, and the square should be aligned carefully for each step. To help ensure accuracy, patented clips can be secured on the square or handscrews can be clamped to the blade.

The length of the rafter being calculated with either of the methods is distance between the centerline of the ridge and the outer edge of the plate. When a ridge board is used, deduct half the thickness of this board from the total length of the rafter before the top cut is made. The overhang is determined by the width of the eaves desired. The width of eaves is determined by the amount of sun blockage that is desired or by a particular eave design. Once the total rafter length has been determined, the rafter is cut to fit and to length.

The top cut on the rafter is the place where the upper end rests against another rafter or the ridge board. This cut is parallel to the centerline of the roof. The bottom, or heel, cut is at the lower end of the rafter, horizontal with the plates. This makes the top and bottom cuts at right angles to each other. By imagining a large square placed on the rafter, it is easier to visualize how the cuts are made. The tongue coincides with the top, plumb, or ridge cut. The blade coincides with the bottom, heel, or plate cut. To obtain the layout of the heel and the plumb cuts, use the 12-inch mark on the blade and the rise per foot of the run on the tongue. The horizontal cut is marked along the blade of the framing square, and the vertical cut is marked along the tongue.

To specifically lay the top and seat cuts the following steps need to be followed. For example if the rafter is 16 feet 6 inches long and the rise is 9 inches per foot of run, the following measurements would be used to mark the layout for cutting. The heel and plumb cuts have already been marked on the rafter using the procedure outlined above. For the bottom or seat cut, lay the square on the rafter so that the 12-inch mark on the blade coincides with the mark that was made for the heel, and the 9-inch mark on the tongue coincides with the rafter edge. Mark along the blade to obtain the line for the seat cut. Move the square to the top end of the rafter. Place it in the same position coinciding with the plumb cut. Deduct half the thickness of the ridge board at the top end. This deduction should be at right angles to the plumb cut. Add tail overhang length to the bottom end. Cut two rafters. Place them on the building to check their accuracy. Use these rafters as patterns for cutting the other rafters.

If a common rafter must be cut for a roof that has a odd number of inches in the span, such as 24 feet 10 inches, use the following procedure. Lay the framing square on the rafter with the 12-inch mark of the blade on the last full layout point and the tongue layout on 9 inches. The run of such a building would be 12 feet 5 inches. The additional inches are added at right angles to the last plumb line after the numbers obtained from the square for each foot of run are measured.

The tail cut may be square, plumb, or a combination of plumb and level. Check the cornice details shown in the architectural plans for exact requirements.

HIP AND VALLEY RAFTERS

A hip rafter is needed when two roof surfaces that slant upward from adjoining walls meet. A valley rafter is needed when two roof surfaces slant downward toward the building. Both rafter types are laid out the same way. The relation of hip and/or valley rafters to common rafters is the same as that of the sides of a right triangle. If the sides forming a right triangle are 12 inches each, the hypotenuse, or side opposite the right angle, is equal to 16.97 inches, usually considered to be 17 inches. A hip or valley rafter can also be thought of as a diagonal of a square if the roof were viewed from above. The square is made up of two common rafters and two sides of the building edge.

Basic Procedures for Laying Out and Installing Hip and Valley Rafters

1. *Hip or valley rafters are laid out in the same way as common rafters except that the 17-inch mark on the blade of the framing square is used instead of the 12-inch mark. The total run of the hip or valley is 17 inches multiplied by the run, in feet, of the common rafter. The lengths of the hip or valley rafters are found on the second line of the rafter table—"length of hip and valley rafters per foot run." Numbers in this table indicate the length of hip and valley rafters per foot of run of common rafters.*

2. *To find the length of a hip or valley rafter, multiply the length given in the table by the number of feet of run of the common rafter. For example, to determine the length of a hip rafter that has a rise of roof that is 8 inches per foot of run (one-third pitch) for a building 10 feet wide requires locating the number along the edge of the square corresponding to the rise of the roof, which is 8. On the second line under this figure is 18.76. This is the length of the hip rafter in inches for each foot of run of common rafter for a one-third pitch. The common rafter has a 5-foot run. Therefore, there are five equal lengths for the hip rafter. The length of the hip rafter is 18.76 inches per 1 foot run. Its total length will be 18.76 × 5 = 93.80 inches. This is 7.80 feet or, for practical purposes, 7 feet 9¾ inches.*

3. To obtain the top and bottom cuts of hip or valley rafters, use 17 inches on the blade and the "rise per foot run" on the tongue. The number 17 on the framing square blade will give the seat cut, and the figure on the framing square tongue will give the vertical, or top, cut.

4. The length of all hip or valley rafters is always measured along the center of the top edge or back. Rafters with overhang are treated like common rafters except that the measuring line is the center of the top edge.

5. The deduction for the ridge is measured like that for the common rafter except that half the diagonal (45 degrees) thickness of the ridge must be used.

6. Hip and valley rafters must also have side or cheek cuts at the point where they meet the ridge. These side cuts are found on the sixth (bottom) line of the rafter table, which is marked "side cut hip or valley—use." The numbers given in this line refer to the graduation marks on the "outside edge of the body." To obtain the side cut for hip or valley rafters, use the number given in the table on the blade of the square and 12 inches on the tongue. Mark the side cut along the tongue where it coincides with the point on the measuring line. For example, to find the side cut for a hip rafter when the roof has 8 inches rise per foot of run (one-third pitch), locate the number 8 on the outside edge of the blade. Under this number in the bottom line is 11½. This number is used on the tongue, and the 12-inch mark is used on the blade. The square is applied to the edge of the back of the hip rafter. The side cut is along the tongue. (See Figure 4-61.) Deduct for half the thickness of the ridge in the same way as for the common rafter; the 45-degree thickness of the ridge stock must be used.

(Figure 4-61)   Using a square

7. In making the seat cut for the hip rafter, an allowance must be made for its top edges. These edges would project above the line of the common and jack rafters if the corners of the hip rafter were not

removed or backed. The hip rafter must be lowered slightly by cutting parallel to the seat cut. This amount varies with the thickness and pitch of the roof.

8. *The 12-inch mark on the square is used in all angle cuts at the top, bottom, and side. The number taken for the fifth or sixth line in the table is the only other one to remember when laying out side or angle cuts. The side cuts are always on the right hand, or tongue side of rafters.*

9. *Additional inches in the run of hip or valley rafters are added in a way similar to that explained for common rafters. The diagonal (45 degrees) is used. Approximately 7¹/₁₆ inches are added for 5 inches of run.*

10. *The tail on a hip rafter must form a nailing surface for the fascia boards. The tail end is cut by finding the center of the rafter and then measuring back from each side of the centerline half the thickness of the rafter. By cutting each of the edges off, a mitered corner is created to which the fascia boards can be nailed.*

11. *Hip and valley rafters are installed after they are laid out and cut. Their installation gives stability and a skeleton to the roof frame.*

JACK RAFTERS

Jack rafters are common rafters that have been "cut off" by the intersection of a hip or valley before reaching the full length from plate to ridge. They lie in the same plane as common rafters and are usually spaced the same and have the same pitch. They also have the same length per foot of run as the common rafters.

Jack rafters rest against the hip or valley rafter. When equally spaced, the second jack must be twice as long as the first one, plus the length of the tail, if any; the third is three times as long as the first plus the length of the tail, if any. This relationship between the larger of the first and following rafters continues for each additional jack rafter. Lengths of jack rafters are given in the third and fourth lines of the rafter tables on the framing square as follows: third line: "difference in length of jacks—16 inch centers"; fourth line: "difference in length of jacks—24 inch centers." Numbers in the table indicate the length of the first, or shortest, jack, which is also the difference in length between the first and second, and between the second and third jack, and so on for each rafter.

Basic Procedures for Laying Out and Cutting Jack Rafters

1. *Find the length of a jack rafter by multiplying the value given in the framing square table by the number indicating the position of the jack. From the obtained length, subtract half the diagonal (45 degrees) thickness of the hip or valley rafter. For example, the length of a second jack with a roof that has a rise of 8 inches to 1 foot of run of the common rafter spaced 16 inches apart can be found by locating (on the outer edge of the blade of the square) the number 8, which corresponds to the rise of the roof. On the third line under this number is the number 19.23. This indicates that the first jack rafter will be 19.23 times 2, or 38.46 inches. For practical purposes, the length is 3 feet 2-½ inches. From this length, deduct half the diagonal thickness of the hip or valley rafter. The same procedure is used when jack rafters are spaced on 24-inch centers.*

2. *Top and bottom cuts on jack rafters are laid out in the following steps. Since jack rafters have the same rise per foot of run as common rafters, the method of obtaining the top and bottom cuts is the same. Use 12 inches on the blade and the rise per foot of run on the tongue. The 12-*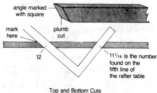

Top and Bottom Cuts for Jack Rafters

(Figure 4-62)   Jack rafters

inch mark locates the line for the seat cut; the figure on the tongue locates the plumb cut. (See Figure 4–62.)

3. *The side cut for a jack rafter is required where the end meets the hip or valley rafter. Side cuts for jacks are found on the fifth line of the rafter tables. It is marked "side cuts of jacks—use." To lay out the side cut, use the number shown in the table on the blade of the framing square and the number 12 on the tongue. Mark along the tongue for the side cut. For example, if a roof has an 8-inch rise per foot of run (⅓ pitch), locate under the 8 on the edge of the square a 10 in the fifth line of the table. Use this 10 on the outside edge of the blade and 12 inches on the tongue. The two numbers will give the line for the required side cut. The side cut should be*

made either with a portable electric saw with an adjustable base and guide or with a radial arm saw. Either of these saws will help to maintain the needed accuracy for the cuts.

4. To assist in laying out jack rafters once the initial layout has been completed, a sliding T-bevel can be set for the side cut. Once the length has been determined, each jack rafter can have the side cut marked by using the T-bevel.

5. Each hip in the roof requires one set of jack rafters made up of matching pairs. A pair consists of two rafters of the same length with the side cuts made in opposite directions.

6. Valley jack rafters are the same as hip jack rafters except the position of the valley rafters is that they angle down to create a valley just as their name implies. The procedure for determining the length of the valley jacks is the same as for hip jacks. The numbers to be used on the framing square are in the fifth line, the same as for hip jacks. The longest valley jack will be the same as a common rafter except for the side cut at the bottom. It is best to start the layout at the building line and work up to the ridge. The side cuts are on the end of the rafter that nails into the valley rafter, and the plumb end is nailed into the ridge. This is just the reverse of the hip rafters.

7. Jack rafters should be erected in pairs to prevent the hip and valley rafters from being pushed out of line.

8. Use 10d nails and space them so they will be near the heel of the side cut as they go from the jack into the hip or valley rafter.

9. As each pair of jacks is installed, look at the valley or hip to determine that it is straight and true. Use temporary bracing to hold the rafters in place until the sheathing is installed. This bracing can be used to move or twist the rafters as needed to bring them into straight alignment. Remove the bracing as the sheathing is being installed.

10. After the nailing has been completed and before the sheathing has been installed, the fascia should be put in place. Fascia boards are nailed to the vertical ends of the rafters. They conceal the rafters, provide a finished appearance, and furnish a surface to which guttering may be attached. Some carpenters prefer to nail 2 × 4s between the rafters before nailing the fascia on. This holds the ends of the rafters secure, especially if the overhang is wide, and gives more support and nailing surface for the 1-inch fascia. The fascia

boards should be beveled on the top edges to match the angle of the roof. The ends should be mitered and carefully fitted.

RIDGE BOARDS

The ridge board is the center piece upon which the rafters rest as they are nailed into place. It is an aid in erecting the roof, because the location of the various rafters is marked on it. The ridge board is made from straight lumber, since it is the spine of the roof structure. Ridge boards for gable roofs are cut to the same length as the building, while ridge boards for hip roofs must be laid out to make allowances 'for the various lumber thicknesses and the methods used to frame the roof. Joints in the ridge should be laid out to occur at the center of the rafter. Ridge boards are laid out by transferring the rafter and ceiling joist layout that has been done on the wall plate to the ridge boards. Once the ridge boards have been cut to length, they should be set across the ceiling joists close to where they will be assembled with the rafters.

When laying out the length of ridge boards on hip roofs, there are a number of factors that can influence the final length. The kind of framing used is one factor, and the way the rafters are brought into the ridge boards is another. For example, when the hip rafters frame directly to the ridge, the ridge must be lengthened by an amount equal to one-half the thickness of the ridge plus one-half the diagonal thickness of the hip rafter. The ridge board must be lengthened one-half the thickness of the common rafter when hip rafters frame in three common rafters. (See Figure 4–63.)

SPECIAL PROBLEMS

Special problems can occur in roof framing where the spans of two roof sections are not equal, preventing the ridges from meeting. To support the ridge of the narrow section, one of the valley rafters is continued to the main ridge. The same method used in the layout of a hip rafter can be used to determine the length of the extended or supporting valley. The valley rafter is shortened at the ridge just like the hip. A single side cut is all that is required to make it ready for fit. The other valley rafter is attached to the supporting valley by means of a square plumb cut.

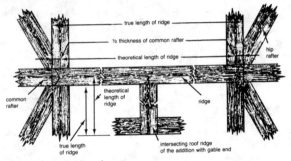

(Figure 4-63) Ridge board

A valley cripple jack is a rafter that is framed between two valley rafters. The side cut is made at an angle on the top that is the reverse of the side cut at the lower end. The valley cripple uses one side of a square for the run. The calculation for the run is equal to twice the distance from the centerline of the valley cripple jack to the intersection of the centerlines of the two valley rafters. The length of the cripple jack is laid out using the same method as that used for a jack rafter. Each end half is shortened by the 45-degree thickness of the valley rafter stock. The side cuts are made in the same way as for regular jack rafters.

Roof openings can create problems in the framing process that require care and attention to the layout process. These openings are generally created to accommodate skylights, chimneys, and other structures. For small openings, the entire framework is first completed, and then the opening is laid out and framed.

For openings in the roof, such as chimney openings, use a plumb line to locate the opening on the rafters from the openings already formed in the ceiling or floor frame. Temporary support strips extending across two additional rafters should be nailed above and below the points where the rafters will be cut. These supports will hold the ends of the cut rafters steady during the process of nailing the headers. Cut the rafters and nail in the headers. Double the headers and add a trimmer rafter to each side if the size of the opening is large.

REROOFING

Reroofing is the process of placing a new roof over an existing surface. The reroofing process may involve laying shingles over an existing roofing surface, or it may require stripping the roof down to the sheathing and installing a new roof as you would on a new structure.

The first step in reroofing is inspecting the existing roof. If the existing roof has 3 or more layers of roofing on it, or if the rafters are sagging or the eaves are rotted under the roofing, the best method of proceeding is to strip the roof back to the frame members and install new sheathing and roofing materials.

If it is determined that the existing roof can be roofed over, then the roof should be inspected to correct any problems. For example, loose shingles should be nailed down. If shingles are warped or curled, they should be nailed down as well. If at the edge of the roof, the shingles and sheathing have rotted, cut away the old shingles and lay in a new board 4 to 6 inches wide and ⅞ inch thick along the edges. This will give a solid nailing base for the new roof.

When reroofing over old wood shingles, it may be necessary to level the existing roof by nailing strips of wood along the bottom or butt edge of the wooden shingles. The wood strips must be the same thickness as the width of the shingles. Old valleys can be built up with wood strips to make them flush with the butts of the wood shingles. Felt then can be laid over hips and ridges and at valleys. (See Figure 4–64.) Extra-long nails should be used to penetrate and hold the new shingles.

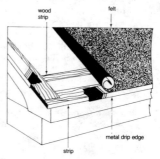

(Figure 4-64) Reshingling

ROOF DECKS

Roof decks are the portion of the roof construction to which the roof covering or roofing is applied and through which the loads on the roof are transmitted to the principal supporting members. Roof decks include sheathing, roof planks or slabs, rafters, purlins, subpurlins, and joists. The principal supporting members include girders, trusses, rigid frames, and the ribs of arches. The surface of the roof deck is expressed in squares (100 square feet).

SHEATHING

Roof sheathing is installed to increase the rigidity and strength of the roof while providing a base for the roof covering. Materials used for sheathing include plywood composites, oriented strand board, waferboard, particleboard, shiplap, and common boards. The most common sheathing used in modern construction is plywood composites, waferboard, or particleboard. The sheathing must be solid if asphalt or other composition shingles are going to be used. For wood shingles, metal sheets, or tile, board sheathing may be spaced according to the course arrangement. The boards should be spliced at the center of rafters, and the longest boards available should be used. For solid sheathing, a careful fit must be made around hips and valleys. The decking must be nailed securely to present a smooth surface upon which the flashing can be secured. Around chimney openings, the sheathing must have a ½-inch clearance.

Basic Procedures for Installing Roof Sheathing

1. *Board sheathing is spliced over the center of rafters, long boards are used, and 8d nails are used to nail the sheathing. Board sheathing is commonly 1 × 4 or 1 × 6, and spacing depends on how much of the shingle is exposed.*

2. *Panel roof sheathing (in 4 by 8 foot panels) is the most commonly used roof sheathing when asphalt shingles are to be laid. Plywood sheathing panels are laid with the face grain perpendicular to the rafters.*

3. *End joints on the panels are spliced in the center of the rafters.*

4. *Small pieces of panel sheathing can be used, but they must be large enough to cover at least two rafter spaces.*

5. *For asphalt shingles, the panel thickness should be at least 5/16 inch if the rafters are set on 16 inch centers. For rafters set on 24 inch*

centers, a ⅜-inch thickness should be used. Slate, tile, and mineral-fiber shingles require ½-inch thicknesses for 16-inch rafter spacing and ⅝ inch for 24-inch spacing.

6. *The panels are nailed to the rafters with 6d nails, spaced 6 inches apart on edges and 12 inches elsewhere.*

7. *If wood shingles are going to be laid and the sheathing is less than ½ inch thick, 1 × 2 nailing strips, spaced according to shingle exposure, should be nailed to the sheathing.*

8. *For a flat deck under built-up roofing, use ½-inch thickness.*

SLOPE AND PITCH

The slope of a roof equals the ratio of the vertical rise to the horizontal run. The ratio can be expressed as *x* distance in 12. For example, a roof that rises at the rate of 2 inches for each foot of run is said to have a 2 in 12 slope. The architectural plans show this information using a triangular symbol above the roof line.

The pitch of the roof equals the ratio of the vertical rise to the span (which is twice the run). For example, if the total roof rise is 2 feet and the total span 12 feet, the pitch is ⅙ (²⁄₁₂, or ⅙).

TRUSSES

Trusses create a framework that is designed to carry a load between two or more supports. They are lightweight and are designed according to accepted engineering practices. They are joined with glue, nails, staple nails, bolts, and special connectors. They eliminate the need for interior load-bearing partitions. In this type of construction, gable ends are usually framed in the conventional manner using a common rafter to which the gable and the studs are nailed. Overhangs at eaves are provided by extending the upper chords of the trusses beyond the wall or by nailing the overhang framing to the upper chords.

Trusses may be constructed on the job site or may be purchased preassembled. Generally the trusses are preassembled and delivered to the site ready for installation. The trusses are set on the wall plates upside down, as needed, the peaks of the rafters are swung upward, where they are secured in place.

If hip and valley construction is necessary, modified trussed rafters can be used. With the aid of computers, the layout of the trusses and the calculation of their load-carrying capability can be completed prior to construction. Trusses are located on 16- or 24-inch centers.

TYPES OF ROOFS

Roofs of buildings can be divided into various types, depending on the shape. There are many different variations on the shapes, but they can be generally grouped into six major categories. (See Figure 4–65.)

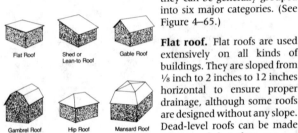

Flat Roof   Shed or Lean-to Roof   Gable Roof

Gambrel Roof   Hip Roof   Mansard Roof

(Figure 4-65) Roof types

Flat roof. Flat roofs are used extensively on all kinds of buildings. They are sloped from $\frac{1}{8}$ inch to 2 inches to 12 inches horizontal to ensure proper drainage, although some roofs are designed without any slope. Dead-level roofs can be made waterproof, but they are not recommended.

Shed roof. A shed roof slopes in only one direction. It may be used over an entire building or in connection with other types of roofs. Most often, a shed roof is used when an addition is made on a structure or when the ceiling is attached directly to the roof frame.

Gable Roof. A gable roof has a center ridge from which the roof slopes in two directions. The two sloping sides form triangular-shaped ends called gables. This simple design is easy to construct and is commonly selected for homes. A gable roof requires that the ends of the structure be framed in since the roof itself stops at the edge of the building.

Gambrel roof. A gambrel roof is a variation of the gable roof. The space under this type of roof can be used efficiently when a long shed dormer is added.

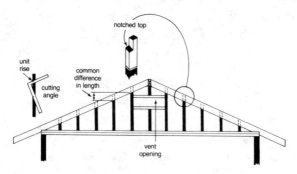

(Figure 4-66)   Laying out a gable

Hip roof. A hip roof slopes in four directions from a central point or ridge. It has the advantage of having overhangs on both the end and side walls. This is one of the most commonly used roof designs for residences.

Mansard roof. A mansard roof has four sloping sides with a deck at the top. This type of roof provides additional space for the rooms on the upper level.

Basic Procedures for Framing in Gable Ends

1. *Mark a squared line across the end wall plate directly below the center of the ridge. Layout for a ventilator is completed by measuring one-half of the opening size on each side of the centerline and marking for the first stud. (See Figure 4–66.)*

2. *Lay out the balance of the stud spacing.*

3. *Place an upright stud at the first space and plumb it with a level. Mark the stud at the point where the underside of the rafter crosses the stud's top edge. Repeat the operation at the second space. The common difference is the distance between the two lengths and can be used to lay out the length of all other studs. By spacing from the centerline, as described, the studs can be cut in pairs.*

4. *The framing square can be used to determine the common difference*

of the stud lengths. Set the square on the stud for the unit run and rise. Mark a line along the blade. Now slide the square along this line until the number for the stud spacing, 16 O.C. for example, aligns with the edge. The distance along the tongue of the square will be the common difference.

5. *Most gable roofs include an extended rake (gable overhang). A gable end frame is required before the roof frame is completed. The rake is created by cutting a set of common rafters for the overhang and then supporting them with lookouts that have been cut to fit in notches in the rafters. The lookouts are 2 × 4s laid flat and extending over two rafters, with their ends nailed into the third rafter.*

6. *Gable ends for brick or stone veneer require that the frame be moved outward to cover the finished wall. By using lookouts and blocking attached to a ledger, a projection can be formed that will cover the finished wall. The ledger for the gable ends should be attached at the same level as that of the ledger used for the top veneer of the cornice construction. This will insure proper alignment on the sides and ends of the building. The support studs can be mounted on this projection and attached to the roof in various ways. Consult the architectural plans for construction details.*

STAIRS

Stairs have taken many forms throughout history, ranging from a log with toe holds hollowed out to ornate spiral staircases that serve as focal points of hotels, mansions, and public buildings. For a number of years, ranch style residences with one floor were built, reducing the need for stairs except for those required to get into and out of the structure. Building styles, however, have changed, and two-story and multilevel homes are now popular. These structures require the use of stairs, landings, and clearance areas, all of which must be carefully planned during construction.

Stairs that are used for movement between levels on a regular basis should be designed with adequate space so that they are comfortable and convenient. Service stairs to attics and basements (unless the basement is used as a main living area) are usually steep and are often constructed of less expensive materials. In some cases, attic stairs come as a kit that can be installed between the ceiling joists as a set-in unit. The stairs fold down for use and fold up out of sight when not in use.

The principal parts of stairs are risers and treads, which are supported by stringers. The risers are cut into the stringers to support the treads. The treads are the cross pieces that are stepped on as the stairs are used. The stringers are long supports that hold the treads and support the load the treads carry. (See Figure 4–67.)

Basic Procedures for Calculating Stairs

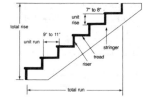

(Figure 4-67)   Parts of stairs

1. *Determine the height, or rise, from the first floor to the next.*

2. *Figure the run, or distance, measuring horizontally.*

3. *Lay out the risers and treads on a stair horse or stringer for a preliminary plan. For example, assume that the total rise is 7 feet 10 inches, or 94 inches. Divide the total rise by 7. (Sometimes 8 rather than 7, is used as the divisor; either number is accurate.) The answer is 13.43. Round the figure off to the closest whole number and divide it into the total rise: 94 ÷ 13 = 7.23, or 7¼ inches. Therefore, the stairs will include 13 risers, each having a height or unit rise of 7¼ inches.*

4. *The number of treads will always be one less than the number of risers. To calculate the total run using a 10½-inch tread, multiply the number of treads by the unit run of one tread. Using the information from the above example, there are 13 risers and 12 treads. The total run, therefore, is 126 inches (12 × 10½), or 10 feet 6 inches.*

5. *There are variations that can be made on stair layouts. The total run can be shortened by using a steeper angle. The number of risers can be decreased and the treads shortened. For example, if the 94-inch total rise is divided by 8, instead of 7, the closest whole number that you will arrive at is 12—the number of risers. Calculate the number of treads (11) and the height of each riser (7.83, or 7-13/16 inches) as*

described above. If the preferred tread is 9 inches, the total run will be 99 inches, or 8 feet 3 inches (9-inch tread width times 11 treads equals the total run).

6. *Tables are available from lumber suppliers for arriving at rise and run and riser and tread ratios.*

STAIR DESIGN

Stairs are designed with two factors in mind: safety and decoration. The first factor is the most important to consider when laying out stairs; the second can be considered after the form and function of the stairs have been established.

The relationship between the riser and tread is most important in stair design. There are three common rules used when calculating the rise-run or riser-tread ratio:

1. The sum of two risers and one tread should be 24 or 25 inches.
2. The sum of one riser and one tread should equal 17 to 18 inches.
3. The height of the riser times the width of the tread should equal between 72 and 75 inches.

Residential structures seldom have treads that are less than 9 inches or more than 12 inches wide. These measurements exclude the nosing. When laying out and constructing stairs, it is important that all of the treads and risers be the same size and of adequate width and height. If the treads are too narrow, the person using the stairs may kick the riser with each step. Also, if the rise-run combination is wrong, using the stairs will cause leg strain and muscle fatigue. An ideal combination for both safety and comfort is a unit rise of 7 to 7-⅝ inches with an appropriate tread width (9-⅜ to 11 inches).

A main stairway should be wide enough to allow two people to pass without contact. There should also be enough width (usually 3 feet) so that furniture can be moved up and down easily. If the main stairway is L- or U-shaped, extra care must be taken in layout to ensure that adequate space has been provided for extra clearance for furniture moving.

Stairs should have a continuous rail along the side for safety and convenience. The rail is generally set at 30 inches above the rake, or slope, of the stairway and 36 inches above landings. (See Figure

4–68.) Stair rails along open stairways have a low partition or banister. On closed stairways, the handrail is fastened to the wall with special metal brackets.

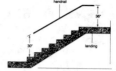

(Figure 4-68)   Banister

The architectural plan should have all of the specifications for the stair layouts. These plans will also have any unusual features that are called for, such as split-level entryways or split landings. The plans generally do not include the exact specifications for the tread mountings, nosing requirements, and the height of the handrail.

STAIRWELL FRAMING

Stairwell framing may be completed during the wall and floor framing process or as one of the last projects in the interior finishing stage. The choice depends on the preference of the carpenter. During either stage of construction, the stairwell must be carefully planned, especially if the architectural plans do not include details for stair framing. Also, local building codes and standards must be followed, as is true with any type of construction.

Providing adequate headroom often is a problem, particularly with small structures. One way of solving this problem is to install an auxiliary header close to the main header. (See Figure 4–69.) This will give a slight extension in the floor area above the stairway. Another method of gaining

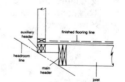

(Figure 4-69)   Stair headroom

headroom is, if the stairway has a closet directly above it, to elevate the floor to give additional space.

If the span on the stairway is greater than 4 feet, the trimmers and headers should be doubled. Headers that are more than 6 feet long should be installed with framing anchors unless they are supported by a beam, post, or partition. Tail joists that are over 12 feet long should be supported by framing anchors or a ledger strip.

STAIRWELL LENGTH

The stairwell length is normally provided in the architectural plans. If this is not the case, the stairwell length can be calculated from the size of the risers and treads. This is done by taking the required headroom, adding to this the thickness of the floor structure, and dividing the sum by the riser height. This will give the number of risers in the opening. When counting down from the top to the tread from which the headroom is measured, there will be the same number of treads as risers. In order to find the total length of the rough opening, multiply the tread width by the number of risers previously determined.

It is helpful to make a scaled drawing of the stairs and floor section to check the calculations. This is a second check to see whether the layout is correct for the required length.

STRINGER LAYOUT

Stringers are the supports upon which the treads and risers rest. There are normally two stringers used in stair construction. If the stairs are extra wide, a third stringer is used for support down the middle of the stairs.

The simplest type of stringer is one made by attaching cleats to 2 × 10s or 2 × 12s on which the treads rest. Another type is made by cutting dadoes into the stock so that the treads fit into the grooves. This method is frequently used for basement construction where no riser enclosure is required.

(Figure 4-70)   Housed stringers

Cutout stringers are used for main or service stairs. In modern construction, prefabricated stairs may be used that include precut stringers, treads, and risers. The housed stringer is another form of stringer construction. The stringer has tapered grooves cut into it. The treads and risers are fitted into these grooves, and then glue-coated wedges are driven into the grooves under the treads and behind the risers. (See Figure 4–70.) The treads and risers are joined with rabbeted edges and grooves or glue blocks. This type of stair construction is dust tight, strong and seldom squeaks. This stair system can be purchased ready for installation, or it can be cut out on the job by using an electric router and a template.

Basic Procedures for Laying Out Open Stringers

1. *Determine the riser height. Do this by taking a story pole (strip of straight lumber) and placing it in a plumb position from the finished floor below through the rough stair opening above. Mark on the pole the height of the top of the finished floor above.*

2. *Set a pair of dividers to the calculated riser height and step off the distances on the story pole. Usually, these distances do not come out right the first time; make the necessary adjustments and try again. Make the adjustments until the spaces on the pole come out even.*

3. *Measure the setting of the dividers. This length will be the exact riser height to use in laying out the stringers.*

4. *For straight stairs use 2 × 10 or 1 × 12 stock that is straight and of sufficient length. Place the stock on sawhorses for layout.*

5. *Beginning at the end that will be the top, place the framing square in the same position as that for laying out rafters. (This procedure is outlined in the section on rafters.) Draw a line along the outside edge of the square blade and tongue. Move the square to the next position and repeat the procedure. There must be a minimum of 4 inches between the cuts made for the stair layout and the back of the stock. Narrower back stock will be too weak to support the stairs when installed.*

6. *Framing-square clips or a strip of wood clamped to the square will ensure accuracy during the layout process.*

Layout for Top End Cuts

cutting line

cutting line

7. *Continue the layout process until the required length of the stringers has been laid out.*

(Figure 4-71) Stringer layout

8. *On the end of the stock that will be the top, extend the last*

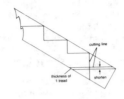

(Figure 4-72)   Cutting stringers

tread and riser line to the back edge. Now mark a cut line back from the top line that is the thickness of the tread. The first calculations did not take into consideration the thickness of the top tread when the height of the riser was being figured. (See Figure 4–71.)

9. *The bottom of the stringer is marked by using an angle that is the same as the rise, but it extends across the stock. When cutting, follow the large X that has been created on the bottom of the stringer; the extra stock is removed, and no other cuts will have to be made. (See Figure 4–72.)*

TREADS AND RISERS

Treads and risers are the support system for stairs that distributes the weight between the stringers. The treads and risers may be left open and if so, must be installed carefully so that there will be no tool marks left as scars. Most often, the treads and risers are not finished but instead are covered with carpeting.

The standard thickness of a main stair tread is generally 1-1/16 or 1-1/8 inches. Softwood or hardwood may be used. Stair treads must have a nosing, which is the part of the tread that overhangs the riser and provides toe room. The width of tread nosing varies from about 1-1/8 to 1-1/2 inches. It should seldom be greater than 1-3/4 inches. A general guide is: as the tread width increases, the nosing width decreases.

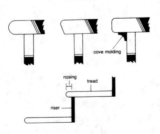

(Figure 4-73)   Nosing types

There are several different nosing designs that can be used. The two most common designs are the rounded edge and the top round with an angle back towards the riser. Cove molding can be added to cover the joint between the riser and tread. (See Figure 4–73.) This molding also covers the nails used to attach the riser to the stringer.

Risers are generally made from ¾-inch stock and should match the wood from which the tread is made. Glue blocks are often fastened to the bottoms of the treads, and the risers are often fastened to the glue blocks. In some construction, the bottoms of the treads have grooves in which the risers are glued. In any case, the risers are fastened to the treads with glue, nails, and/or screws.

Some stairs have open risers, which means that no riser boards are installed. The open risers are used to achieve an open effect or to save construction time and costs, particularly for stairs that are not a main part of the structure.

TYPES OF STAIRS

The primary function of stairs is to allow passage from one level of a structure to another. However, stairs can also be used as an integral part of the decor. The final selection of a stair design depends on personal preference, cost, and available space.

OPEN STAIRS

Open stairs are generally constructed as main stairways that are open on one or both sides. They require some type of decorative enclosure and support for a handrail. Typical designs for handrails involve an assembly called a balustrade. The principal members of a balustrade are newels, balusters, and rails. They are generally manufactured and then brought to the job and assembled. Most often, factory-made staircase parts are used because they can be installed quickly and easily. A wide range of designs are available for staircase construction, and a lumber supplier can provide information on design and construction features.

SPIRAL STAIRS

Spiral stairs are generally made from metal and come in a kit ready for installation. They can be designed with up to 30 steps to fit areas that are up to 22½ feet high. The National Uniform Building Code permits use of spiral stairways for exits in private dwellings or in some other situations where the area served does not exceed 400 square feet.

WINDER STAIRS

Winder stairs are not frequently used in modern construction because of the difficulty of creating adequate tread width. This type of stair system is encountered most often in remodeling homes that were built before the turn of the century.

Winder stairs have a severe turn in a part of the stair course. Care must be taken in developing the needed tread width along the line of travel. The winder tread width should be the same as the tread width used in the straight run, but this is often very hard to maintain because of space restrictions. If a winder stair system is being used, a full-size or carefully scaled layout of the stairs should be created to ensure that the best pathway for the line of travel has been determined.

TRIM AND MOLDING

Interior trim and molding is installed as decoration and to conceal spaces between door frames, window frames, and floor coverings. It is the last stage in the construction process, completed after all the windows have been set, the doors have been hung, and the floor coverings have been laid. Depending on the finish specifications, the trim and molding may be softwood or hardwood, vinyl covered, painted, or clear finished. Trim is normally used in trimming out windows and doors; molding is used in trimming around rooms between the floor covering and the wall surface.

Exterior trim and molding is generally a part of the prehung door or window and requires no trim work. Since the trim is attached to the door or window frame, it sets in place as part of the unit and requires only surface finish to make it complete.

BASEBOARD MOLDING

Baseboard runs around the entire room and is fitted after door casings and cabinets have been installed. It generally comes in 16-foot lengths, so spliced joints are not often needed in residential settings. If the baseboard must be spliced, use a mitered lap joint, called a scarf joint. Position the joint so that it will lap at a stud and can be nailed there.

Basic Procedures for Installing Baseboard Molding

1. *The first piece of baseboard should be cut so that it makes a tight butt joint with the intersecting wall surface.*

2. *The second piece of baseboard running from the internal corner is joined to the first piece with a coped joint. This joint is formed by first cutting an inside miter on the end of the baseboard. Then using a coping saw cut along the line where the sawed surface of the miter joins the curved surface of the baseboard. Make the cut perpendicular to the back face. This forms an end profile that will match the face of the baseboard. (See Figure 4–74.)*

3. *All outside corners are joined with a miter joint.*

4. *Before nailing the baseboard into place, check the joints for final fit.*

5. *To nail the baseboard into place, hold the board tightly against the floor and nail it in place with finishing nails that are long enough to penetrate well into the studs. Set the nails as you would for any finish work.*

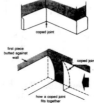

(Figure 4-74)   A coped joint

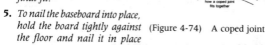

6. *Baseboards are normally butted up against the door casings. Casing material is designed to accommodate the slightly thinner dimension of the baseboard.*

BASESHOE MOLDING

Baseshoe molding is installed along the bottom edge of baseboard. Whether baseshoe is used at all is a matter of personal preference.

Baseshoe molding is laid out and cut the same way as baseboard molding. Inside joints are coped, and outside joints are mitered.

When baseshoe ends at a door casing, a miter cut is made at the end rather than leaving the end with a straight cut. The mitered end is attractive, and the removal of the excess stock helps to prevent the baseshoe edge from being hit by a person walking through the door. In this case, safety and appearance are combined in the finish joint.

FACTORS TO CONSIDER

- Make sure you know where the studs are as you prepare to nail the baseboard. Once the studs are found, you may want to drive nails in the wall to outline the specific points where you want to nail the baseboard. This is especially true if a scarf joint is involved. Be sure to drive the test nails below the width of the baseboard so that the nail holes will not show.

- The coped joint takes longer to cut for inside joints, but the end results are well worth the time and effort. The joints fit better and give a smooth, finished appearance. A straight mitered inside joint will pull apart when nailed, and if the wood shrinks or the building shifts, the mitered joint will gap. A coped joint accommodates these construction problems.

INTERIOR WINDOW TRIM

Window trim consists of four parts: casing, stool, apron, and stops. (See Figure 4–75.) These parts come precut; all you have to do is install them, or you may have to cut them to length for fit.

Basic Procedures for Installing Interior Window Trim

1. *To mark the stool for cutting, hold it level with the sill and mark the inside edges of the side jambs. Also, mark a line on the face where the stool will fit against the wall surface.*

2. *Cut out the ends and check the fit. Open the lower sash slightly to slide the stool into position. Bring the sash down carefully on top of the stool and draw the cutoff line so that the sash will clear the stool. Allow about $1/16$ inch between the front edge of the stool and the window sash.*

3. *Position a piece of side casing and measure beyond it about ¾ inch. Mark the cutoff lines for the ends of the stool.*

4. *Cut the ends to length, sand the surfaces, and nail the stool into place. Some installations require that the stool be bedded in caulking compound.*

5. *Next install the casing. Set a length of casing on the stool and mark the position of the miter on the inside edge. Note that some carpenters prefer to set the casing back from the jamb edge about ⅛ inch.*

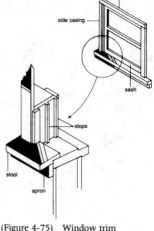

(Figure 4-75)   Window trim

6. *Cut the miter.*

7. *If hardwood is being used, drill pilot holes in the casing.*

8. *Nail the casing in place and repeat the procedure for the other side.*

9. *Lay out the head casing by cutting a miter on one end of the casing stock. Place the head casing in position and mark the other miter cut on the inside edge.*

10. *Nail the head casing in place.*

11. *The last step is to cut the apron. The length of the apron is equal to the distance between the outside edges of the side casing. After cutting the apron to length, sand the surfaces that will be exposed and nail the apron in place.*

12. *Some windows do not use stool and apron trim. In place of this trim, a beveled sill liner is installed along with the standard casing. This trim is called picture frame trim.*

13. *When the desired fit has been achieved for all of the trim members, complete the nailing process. Use a nailset to set the finish nails below the surface. These nail holes will be filled as a step in the finishing process.*

WALLS

Walls serve a number of functions. They support the ceiling and roof of a structure, keep the elements out, and provide a pleasing interior. Wall construction has both exterior and interior applications; the former is more for protection, while the latter is more for finish. This section deals with the framing, covering, and finishing of exteriors. Interior walls are discussed in the section on the installation of wall and ceiling coverings.

CORNICES

Cornices, or eaves, are formed by the roof overhang where the roof and wall meets. It is critical to make a proper fit. (See Figure 4–76.) Cornice construction is very time consuming, so pre-fabricated materials are used to speed the process. There are a number of available systems, and a lumber supplier can provide information on the full range that can be used in a given setting. One common system employs ⅜-inch laminated wood-fiber panels. These are factory primed and are available in a variety of widths from 12 to 48 inches. The panel can be equipped with factory-applied screened vents.

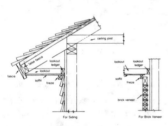

(Figure 4-76)   Parts of a cornice

The amount of overhang is established by the architectural plans. The rake is the part of a roof that overhangs a gable. The cornice and rake are considered a part of the exterior trim package. The rake is finished to match the design of the cornice, and similar procedures are used to install both of them.

Basic Procedures for Constructing Cornices

1. *Install the fascia boards on the rafter ends at the time the roof sheathing is installed. The method for completing the fascia-board*

installation is described in the roofing section. The fascia boards must be straight, true, and level with well-fitting joints so that the cornice work can be properly completed.

2. *Normally the fascia boards are nailed directly to the rafter tails or ends. A 2 × 4 ribbon may be nailed on first to align the tails of the rafters. The fascia is then nailed to this ribbon. The corners of fascia boards should be mitered to ensure a tight joint and a smooth surface without ends showing.*

3. *A cornice can be framed with a horizontal soffit by first nailing a ledger strip along the wall if lookouts are going to be used. This is done by using a carpenter's level to locate points on the wall that are level with the bottom edge of the rafter. (See Figure 4–77.) Then snap a chalk line where the ledger is to go.*

mark for lower edge of ledger board

level

(Figure 4-77)   Soffit

4. *Lookouts are generally made from 2 × 4s. They are positioned at each rafter. If the soffit material is rigid, the lookouts can be placed on every other rafter. The lookouts are toenailed into the ledger strip and nailed through to the rafters.*

5. *If the soffit material is thin and flexible, a nailing strip should be positioned between the rafters. This strip will give additional support to the soffit material.*

6. *Some builders like to cut a groove in the back of the fascia board to receive the soffit material. This is optional, depending on the soffit material used.*

7. *The soffit material is installed after the frame is completed. A common soffit material is ⅜-inch good one-side plywood. The soffit material is cut to size and then secured to the frame with rust-resistant nails or screws.*

8. *Sloping cornices are installed in the same way as straight cornices except that the soffit material is nailed to the underside of the rafters. A frieze board may be attached to the rafter first to help hold one edge while nailing the soffit.*

9. *Soffits should have screened vents installed throughout their length in order to permit the flow of air through the enclosed cornice section into the attic space. The screen vents come in several different sizes and should be selected based on the size of the cornice and the amount of airflow required.*

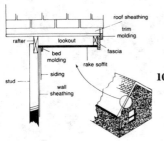

(Figure 4-78) Roof rake

10. *The rake is constructed in the same manner as the cornice. A nailing strip is attached on the siding of the structure and another is attached on the inside of the outer edge of the rake. The soffit material is then attached to the ledgers with rust-resistant nails or screws. (See Figure 4–78.)*

11. *If you use prefabricated cornice materials such as laminated wood-fiber panels, leave space between the ends for expansion. Use 4d rust-resistant nails, spaced about 6 inches apart, along edges and intermediate supports. Nail each panel from the end that has been butted up against the other. Nail the end first and then nail along the edges. The nails should be driven so that the underside of the nailheads are flush with the panel. Care must be taken not to mar the panel or drive the nails too deeply, which will pull the panel and create an uneven surface.*

12. *One cornice system uses 20-gauge steel channels that attach to the back side of the soffit material. The steel channels have prongs that are forced into the soffit material by inserting a 2 × 4 stock piece into the channels, positioning them on the soffit material and then driving them into position by striking the 2 × 4. If this system is used the lookouts are eliminated, since the channels provide the required support. The soffit material is nailed into place along the front and back edges. A special H-clip is used to hold and join the end joints in place.*

13. *Metal soffits can be used and are installed by first snapping a chalk line along the side wall. The chalk line is laid out level with the bottom edge of the fascia board. A wall hanger strip is nailed in*

place. The soffit panels (normally 9 inches wide) are inserted in the wall hanger strip and nailed in place on the bottom edge of the fascia board. When the panels are all in place, cut and install a fascia cover. The bottom of the cover is hooked over the end of the soffit panels and nailed into place through prepunched slots located along the top edge. Check the manufacturer's directions for the exact installation specifications.

14. *Gaps or irregularities between the soffit material and the structure may be covered by moldings. The use of moldings should be kept at a minimum in order to maintain a smooth, trim appearance.*

CORNICE DESIGNS

Cornice designs are created by the amount of overhang formed by the roof and architectural style desired. Roof overhangs range from very narrow to extra wide. Extra-wide overhangs are used particularly if solar factors have been considered in the design. Also, bay windows require wide overhangs in order for the cornice and the top of the bay window to meet properly.

There are two basic cornice designs: the open cornice and the closed or boxed cornice. The open cornice is used to expose rafters and the underside of the roof sheathing. This style is often used on recreational or seasonal homes and log cabin homes. The closed cornice has soffit material installed under the rafters between the fascia board and the side wall. There are a number of variations on the closed design that produce different finish effects. The architectural plans detail how the closed cornice is to be completed in order to fit a particular style.

FRAMING WALLS

BRACING

To counter lateral stress, exterior walls usually need some type of bracing. The bracing is normally of two types. The let-in form of bracing uses a 1 × 4 piece of lumber placed at a diagonal run from the top and sole plate. Where the brace rests on the studs, the studs are notched to receive or "let in" the brace. Care must be taken to make the diagonal angle such that the brace does not interfere

(Figure 4-79)   Wall bracing

with window or door openings. (See Figure 4–79) Another form of bracing is a 2-inch wide metal strap made from 18- or 20-gauge galvanized steel. The studs are notched to receive the band, and it is nailed in place after the wall frame has been erected and plumbed.

When plywood sheathing is used on the corners of the structure, the need for diagonal bracing is eliminated. The sheathing is installed after the frame has been plumbed and is ready to be stabilized.

CORNERS

There are several different methods that can be used to form outside corners of the wall frame. Corners are formed when a side wall and an end wall are joined. In the process of joining these two

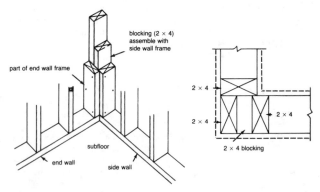

(Figure 4-80)   Corner frame

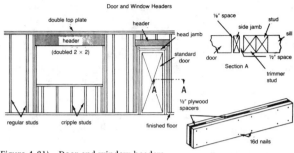

(Figure 4-81)   Door and window headers

walls, adequate strength must be provided as well as nail surfaces for the interior wall covering. One method of creating a corner is to nail two 2 × 4s together with blocks between and a 2 × 4 flush with one edge. In platform construction, the wall frame is usually constructed on the subfloor and then raised into place. The corner is formed with adequate nailing surface as the walls are brought together. Another method of creating a corner is to nail a 2 × 4 to the edge of another 2 × 4, with the edge of one flush with the side of the other. (See Figure 4–80.) Corners for platform construction can be built separately. They are set in place, plumbed, and braced. The wall sections are then built and set in place around the corners.

HEADER CONSTRUCTION

Header construction is required when rough openings are created for doors, windows, or other openings in the walls of the structure. The headers distribute the weight of the roof and ceiling across the rough openings. The size and length of the headers are determined by the size of the rough opening. The rough opening's size is specified in the plans. Headers are made by nailing two pieces of 2× material (2 by standard dimension lumber) together with a ½-inch spacer between to create the same thickness as that of the wall. (See Figure 4–81.) The exact length of the header will be equal to the rough opening plus two trimmers (3 inches). The width of the lumber used in the header depends on the width of the opening. Specific header requirements are often provided in local building codes. If the load is especially heavy or the span unusually wide, a truss may be used. If a truss is called for, it will be outlined in the plans. An extra stud is normally installed around the rough opening for additional support. Also, the extra studs provide nailing surfaces for the window and door casings. Many builders double the rough sill to provide a nailing surface for the window stools and aprons.

When the spans are unusually wide, such as with bay windows, headers and cripple studs can be put in by cutting them to fit the space between the top of the header and the upper plate. In order to save time and provide more support, you can select lumber for the header that is the same width as the distance from the top of the rough opening to the top plate.

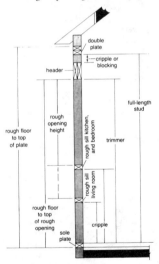

(Figure 4-82) Master stud

MASTER STUD LAYOUT

Our way of laying out stud positions on walls is to make a master stud layout pattern. The layout can be made on a straight 1 × 4 or 2 × 4. First lay out the distance from the rough floor to the ceiling. This dimension can be taken from the plans. Next lay out the header. If several headers are used, they can be marked on top of each other. Measuring down from the bottom side of the header, lay out the height of the rough openings. The rough sill is marked in next. From this full-size layout, the length of the various studs (regular, trimmer, and cripple) can now be marked. (See Figure 4–82.) The other side of the pattern should be used to mark door openings that have different header heights or for windows. This will help to avoid confusion. A separate master stud layout will be required for each level of a multi-story or split-level house.

Basic Procedures for Constructing Wall Sections

1. *Using the master stud layout pattern, cut the various stud lengths. The full-length studs come cut to length, so they are ready for assembly.*

2. *Cut and assemble the headers. Their lengths and the length of the rough sill can be taken directly from the plate layout.*

3. *To assemble the wall sections, move the top plate about a stud's length away from the sole plate. With the layout markers facing inward, turn both plates on edge. At each position marked on the plates, place a full length stud, crown up. The studs are then nailed to the top and sole plates, using two 16d nails through each end of each stud.*

4. *The trimmer studs should be set in place and nailed to the full-length studs. Place the header so that it is tight against the end of the trimmer and nail through the full-length stud into the header using 16d nails.*

5. *Cripple studs for window openings can be positioned by transferring the markings from the sole plate to the rough sill. Some builders wait until they erect the wall section before installing the lower cripples and the rough sill. If you choose to do this, toenail the lower ends of the cripple studs to the sole plate.*

6. *Install any bracing that is required, and add studs or blocking at positions where partitions will intersect outside walls.*

7. *Many times, the wall sheathing is applied before the wall section is erected. If you choose to do this, the wall section must be square. Check for square by making diagonal measurements across the corners. To keep the wall section square put a temporary brace across one corner, or nail two edges of the frame temporarily to the floor.*

8. *Often the wall sections are nailed together with a power nailer. If you are using a power nailer, be sure that it is set correctly to drive the nails completely into the lumber, and wear eye protection when operating the nailer.*

9. *The wall sections are raised by hand, wall jacks, or fork lift. If you are raising the wall sections by hand, remember that the length of*

the section should be limited according to the strength you have to raise it. If you are using power equipment or wall jacks, the length of a section is not as important a consideration. Before raising a wall section, apply a temporary bracing diagonally to keep the wall plumb. Also, temporary blocking should be placed on the edge of the floor frame to prevent the wall section from sliding off the platform. Once the wall section is erect, it should be secured with braces running from the top of the wall to the subfloor at about a 45-degree angle. After making necessary final adjustments in the position of the sole plate, nail it in place using 20d nails driven through the subfloor and into the joists.

10. *Bring the wall section into plumb at the corners and midpoint of the wall. The braces can be moved back and forth as needed to accomplish this. A plumb line can be used as the measure, but in most cases a carpenter's level is sufficient.*

PARTITION INTERSECTION

Partition walls divide the inside space of a building. These walls, in most cases, are framed as a part of the building. Where floors are to be installed after the outside of the building is completed, the partition walls are left unframed.

There are two types of partition walls: load-bearing and nonload-bearing. The bearing type supports ceiling joists. The nonbearing type supports only itself. This latter type can be put in any time after the other framework is installed. Only one cap or plate is used. A sole plate should be used in every case, as it helps to distribute the load over a larger area.

Partition walls are framed the same as outside walls, and the door openings are framed the same as exterior door openings. Where partitions meet outside walls, it is essential that they be solidly fastened. This requires extra framing. The framing must be arranged so that the inside wall coverings provide a nailing surface for finish wall covering materials. There are several methods whereby the extra support needed can be supplied as well as creating a nailing surface for the wall covering materials.

1. Install extra studs in the outside wall. Attach the partition to them.
2. Insert blocking and nailers between the regular studs.

3. Use blocking between the regular studs and attach patented backup clips to support inside wall coverings at the inside corners.

PLATE LAYOUT

The top plate ties the studding together at the top and forms a finish for the walls. It also furnishes a support for the lower ends of the rafters. It is the tie between the walls and the roof. The top plate is made up of one or two pieces of lumber that are the same width as the studs. Where the plate is doubled, the first plate or bottom section is nailed with 16d or 20d nails to the top of the corner posts and to the studs. The corners are constructed by butting together the wall sections with single plates. Next the second plate of one section is nailed into place so that it overlaps the single plate of the adjoining wall section. The second plate is nailed into place using a 16d or 20d nail over each stud, or the nails can be spaced so that there are two nails every 2 feet. The edges of the top plate should be flush with those of the bottom plate, and the corner joints should be lapped.

The sole plate carries the bottom end of the studs and consists of a single plate nailed to each stud with two 16d or 20d nails. If the wall section is assembled before it is set in place, the nails go through the sole plate into the end of the studs. If the sole plate is set in place first, then the studs are toenailed into the sole plate. The sole plate is nailed through the subfloor and into the joists for maximum holding power.

When laying out the plates, select straight lumber, so that you won't have to waste time straightening the walls later on. The layout of plates is the same whether for 2 × 4s or 2 × 6s except for the distance between studs. Check the plans for the exact width between studs (2 × 4s are generally set 16 O.C., and 2 × 6s are generally set 24 O.C.). The commonly used widths are 16 or 24 inches O.C. Lay the plates (for stud layout) side by side along the location of the outside wall. The length of the plates is determined by the amount of weight that can be lifted into place. If wall jacks or a forklift is going to be used, the wall sections can be much longer than if they are going to be set in place by hand.

The plates along the main side walls should be laid out first. The ends of the plates should be aligned with the floor frame and then marked at the regular stud spacing all the way along both plates. You might want to tack the plates to the subfloor to hold them in place while doing the layout. Review the architectural plans to determine the exact location of all of the rough openings. Mark a

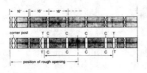

(Figure 4-83)   Stud layout

centerline for each door and window opening on the plates. Measure off one-half the width of the opening on each side of the centerline. Mark the plate for trimmer studs outside of these points. On each side of a trimmer stud, include marks for a full length stud. Identify

positions with the letter *T* for trimmer studs and *X* for full length studs. Now mark all of the stud spaces located between the trimmers with the letter *C*—this designates them as cripple studs. (See Figure 4–83.)

Where intersecting partitions butt, lay out the centerlines. Corners must be laid out very carefully so that they will fit correctly when the wall sections are brought together.

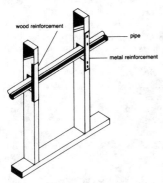

(Figure 4-84)   Frame with plumbing

PLUMBING IN WALLS

Where plumbing is run through walls, special construction may be required. Depending on the size of the drain and vent pipes, a partition may have to be made wider. Usually a 2 × 6 wall will accommodate the required drain and vent widths. When the pipes are run horizontally, holes are drilled in the studs or they are notched out. A metal strap can be attached to the notched stud to bridge the notch and strengthen the stud. (See Figure 4–84.) The strap also protects the pipe from accidental damage should a nail be driven into the stud at that point. Similar protection should be provided for electrical wiring in walls.

ROUGH OPENINGS

Rough openings (R.O.) are the areas in which doors and windows will be set. The plans for the structure will detail the exact size of the rough openings. These plans will also specify the location of these rough openings. Normally the measurements for the rough openings are taken from corners and/or intersecting partitions to the centerlines of the openings.

The height of the rough openings can be taken from the elevations and sectional views. Their size will be shown on the plan view or listed in a table called the door and window schedule. The size of the rough openings is always listed with the width given first and the height given second. Careful study of the plans is absolutely necessary to ensure that the rough opening layouts are correct the first time—saving you much time and effort in the long run.

STORY POLE

A story pole can save you much time when laying out wall sections. A story pole is a long measuring stick upon which marks have been made for every horizontal member of the wall frame. (See Figure 4–85.) By making a story pole for the job, you will not have to go back to the plans every time you lay out a new wall section. Care in marking the height of the horizontal members on the story pole will not only save you time but also maintain accuracy. The story pole is a must when working on split-level homes, since you will be required to use many different stud lengths and heights.

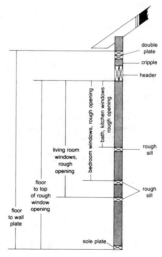

(Figure 4-85)   Story pole

SPECIAL FRAMING

Special framing may be required within a standard structure or within structures that have special design features. Framing for special design features can be accomplished by carefully following

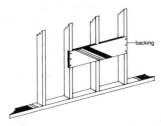

(Figure 4-86a)   Using backing to provide support

the details of the architectural plan. In some framing situations, however, the plans do not include the details. An example of this is framing for an atrium. You will have to not only visualize the construction details but also lay out and construct the floor frame to carry the projection and build the wall and roof frame to support it. If you understand the basics of standard framing, you generally will not experience too much difficulty in applying these principles to the special framing problem you face.

Trilevel or split-level homes also present special problems. Generally, a platform type of construction is used for these homes. There are several ways that the walls for these homes can be framed. A well-laid-out and accurate story pole is extremely useful in working out the framing problems of split-level structures. The

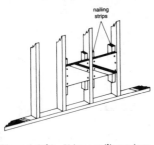

(Figure 4-86b)   Using a nailing strip to provide support

plans should prescribe the type of construction and the distances between floor levels; if not, consult with the architect.

Another type of framing problem occurs where heavy fixtures need wall support. An example of this is the framing required to set a bathtub. Extra joists must be put in the floor and blocks attached to each stud to help distribute the weight of the tub more evenly. If wall-mounted fixtures are called for, extra support can be given to the area in which they will be mounted by cutting spaces into the studs and filling in backing. A key rule here is never cut back more than 25 percent of the stud width on bearing walls or partitions. Another method of giving support is to nail strips to the studs and then nail blocking to the strips. (See Figure 4–86 (a) + (b).) Again, the plans provide the exact location of where the backing needs to be installed.

PAINTING AND MAINTENANCE

Wood or fiber siding is subject to decay and weathering. By properly protecting the surface of the siding, the effects of decay and weathering can be reduced or eliminated. Siding comes either primed or unprimed. If the siding does not have a factory-applied primer coat on it, then one needs to be applied within a few days after installation. The paint used for primer is usually an oil-base paint that creates a mildew-proof and moisture-proof barrier until the surface coat can be applied. If the siding comes with a primer coat, then a surface coat should be applied when the exterior trim is finished. Commonly used surface paints are latex based and are available in a wide variety of colors and finishes.

SIDING

Siding comes in a variety of different materials, styles, and forms. The following section covers commonly used siding.

ALUMINUM

Aluminum siding is a low-maintenance, weather-resistant siding that comes with a baked-on enamel finish. It is made to resemble wood siding in the surface texture. Aluminum siding is made for use on new or existing construction. This siding can be applied over wood, stucco, concrete block, and other surfaces that are structurally sound. The basic specifications (gauge and alloy) for aluminum are set by the Aluminum Siding Association. The siding comes in a variety of horizontal and vertical panel styles. Both smooth and textured designs are produced with varying shadow lines and size of face exposed to the weather. Some siding is produced with an insulated back, composed of impregnated fiberboard.

Manufacturers supply directions for the installation of their products. These should be carefully followed since there are installation differences among the different aluminum siding manufacturers. The panels come with prepunched nail and vent holes. The panels are cut to length, and special corners and trim members are formed on the job to fit the unique requirements of the structure that is being sided. Grounding must be a part of the installation process for aluminum siding to prevent energizing the siding and creating an electrical hazard. The Aluminum Siding Association recommends that a No. 8, or larger wire be connected to any convenient point on the siding and to the cold water service or the electrical service ground. The connecters should be UL (Underwriters Laboratories) approved.

HARDBOARD SIDING

Hardboard siding has been improving in quality, durability, and application over the past few years, and as a result, it is a very popular form of siding in construction. Hardboard siding comes in either vertical or horizontal styles. This siding expands more than plywood siding, so allowance must be made for this expansion. The manufacturer's recommendation for the expansion joint is ⅛ inch. Hardboard siding panels are available in a standard width of 4 feet and lengths of 8, 9, and 10 feet. Lap siding units are usually 12 inches wide by 16 feet long. However, narrower widths can be purchased as needed. The most common thickness is ⁷⁄₁₆ inch. Hardboard siding comes in a wide variety of textures and surface treatments. Most of the panels have a factory-applied primer coat. Prefinished units are available, and these come with matching batten strips and trim members.

When installing hardboard siding, follow standard procedures described by the manufacturer. This siding should be installed where the studs are 16 O.C.; 24 inch centers allow the siding to flex causing a wavy appearance. The nailing base must be firm and adequate to prevent the siding from pulling away from the sheathing. The nails are generally 8d galvanized coated and should be located at least ½ inch in from the edges and ends. The joints should be sealed with a high-quality caulking compound.

Basic Procedures for Installing Horizontal Hardboard Siding

1. *A story pole is a great aid for installing horizontal siding, and it should be prepared as a first step. To lay out a story pole, lay out the distance from the soffit to about 1 inch below the top of the foundation. Divide this distance into spaces equal to the width of the siding minus the lap. Adjust the lap allowance (maintain minimum requirements) so that the spaces are equal. When possible, adjust the spacing so that single pieces of siding will run continuously above and below windows or other wall openings without notching. When the layout is complete, mark the position of the top of each siding board on the story pole.*

2. *Hold the story pole in position at each inside and outside corner of the structure and transfer the layout to the wall. Also, make the layout along the window and door casings. You may want to set nails at these layout points, since it will be easy to attach lines for chalking if they are in place, or they can be used as guides for aligning the siding.*

3. *Nail a beveled spacer strip of 1 × 3 lumber along the foundation line. This will provide the proper tilt for the first course. (See Figure 4–87.)*

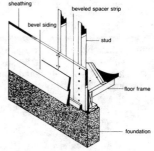

4. *Apply the first piece, allowing the butt edge to extend below the strip to form a drip edge.*

5. *Inside corners can be formed with a square length of wood or metal with the siding butting up against it.*

6. *Outside corners are almost al-* (Figure 4-87)   Siding
ways made from prefabricated metal corners that slip under the course directly above the corner being covered and are nailed in place. The corners come in sizes matching the widths of different siding.

7. *The siding needs to be cut so that it fits tightly against window and door casings, corner boards, and adjoining boards (depending on the joint spacing called for by the manufacturer). When cutting siding, use a smooth-toothed blade, and when the cut is completed, smooth the ends of the siding with a few strokes of a block plane.*

8. *Stagger the joints of the siding so that they do not align directly above one another. This will prevent the formation of a water channel, which can lead to erosion and seapage.*

9. *Nail the siding in place using 8d zinc-coated steel, aluminum, or other noncorrosive nails. The nails will hold best if they have spiral grooves on the shanks. Drive the nails flush with the siding and ½ inch from the edges and end.*

MINERAL-FIBER SIDING

Mineral-fiber siding is made from the same materials as mineral-fiber roof shingles. This siding is usually prefinished with long-lasting factory-baked coatings. It is available in a variety of textures and colors. It is also available with either straight or wavy exposed butt lines. This siding form is not used as much in construction today because of the improvements in and reduced costs of other siding materials such as vinyl and aluminum.

Mineral-fiber siding comes in units (called shingles) 24 inches wide and 12 inches deep. When applied, the shingles are lapped 1½ inches at the head, leaving an exposed area of 10½ by 24 inches. The siding is sold in squares. (A square covers 100 square feet of wall surface.) Ordinarily, there are three bundles per square.

The siding is handled, stored, and cut the same way as mineral-fiber shingles. The layout of the wall surface and the treatment of the corners is the same as for other siding materials. Nailing strips must be put up for the siding, since composition siding will not hold the nails and the siding units. The nailing strips should be placed to overlay the top of the lower course by about ¾ inch. The ends of the nailing strips must be positioned over studs. Vertical joints between the shingle units must not occur over nailing strip joints. The mineral-fiber siding unit is applied with the bottom edge overlapping the wood strip ¼ inch to provide a drip edge.

If the sheathing consists of lumber, plywood, or other materials that will serve as an adequate nailing base, the application of the siding can be made with the aid of a shingle backer. The shingle backer produces an attractive shadow line along the lower edge of each course of siding. If the backer method is used, first nail the backer in place with 1¼ inch galvanized nails spaced about 3 inches from the top edge. If this is carefully positioned along the chalk line, the siding units can be quickly lined up along the top edge.

PLYWOOD SIDING

Plywood siding has many applications beyond that of straight siding, and thus its appeal as a siding material is great. It can be used to provide vertical treatment to gable ends, fill in panels above and below windows, and establish a continuous decorative band at various levels along an entire wall. All plywood siding must be made from an exterior type of plywood. The panel sizes are 48 inches wide and 8, 9, and 10 feet long. A $3/8$-inch thickness is normally used for direct application to studs. A $5/16$ inch thickness may be used over an approved sheathing. Thicker panels are used if the surface texture consists of deep cuts.

The panels are generally installed vertically, eliminating horizontal joints. The vertical joints should be positioned over studs. The panels should be nailed at not more than 4 inches O.C. at the vertical joints, and the nails should penetrate the studs at least 1 inch. The vertical joints can be treated in a number of ways; for example, they can be butted, lapped, or covered with batten strips.

SELECTED SIDING CHOICES

There are a number of siding choices available that provide alternative surface finishes, ease of installation, or cost savings. Vinyl siding, made from polyvinyl chloride, has been gaining in usage in the past few years. It is about $1/20$ inch thick and comes in either horizontal or vertical units and in various widths up to 8 inch. Vinyl siding is installed much like aluminum siding and uses a backer board. The panels are designed with interlocking joints to create a moisture-proof barrier. The manufacturers provide complete instructions for installation. The biggest advantage of this form of siding is its long, maintenance-free life.

Stucco is a commonly used exterior finish in many parts of the country, primarily in the southwest. The finish coat of stucco can be tinted by adding coloring, or the surface can be painted. Wood sheathing, sheathing paper, or metal lath is used to create the base for stucco. The metal lath must be rustproof and spaced at least $1/2$ inch away from the sheathing. This spacing allows the base coat (called the scratch coat) to flow through the mesh and be thoroughly embedded in the lath. Metal or wood molding with a groove that "keys" the stucco is applied at building edges. This type of molding is also used to frame openings to form a strong, smooth edge for the stucco. Stucco application requires experience in using the material and creating the type of surface texture desired.

BRICK OR STONE VENEER

When brick or stone is used as ornamental siding instead of the actual wall, it is called a frame wall. The building plans must specify how far the foundation must extend in order to accommodate the veneer. The veneer must be set away from the structure to provide 1 inch of air space. Base flashing must be placed from the outside face of the wall over the top of the ledge and 6 inches behind the sheathing.

The brick or other veneer selected should be suitable for local use. The quality of brick and stone varies greatly, so it is best to obtain the materials from a reputable local dealer.

VERTICAL SIDING

Vertical siding comes in many different patterns and surface textures. It can be made to appear as rough-cut lumber or batten-board finish.

The batten-board effect can be achieved by using 4 × 8 foot or 10 foot composition or plywood sheeting and covering the joints with vertical strips of 1 × 2 inch lumber.

Vertical siding is used frequently to enclose the gable ends of roofs. The sheets are large enough to allow for quick installation but are light enough for easy handling.

Vertical siding is also made from solid wood that is 6 to 8 inches wide. This form of siding is most frequently used when a rustic appearance is desired.

WOOD SHINGLES

Wood shingles are used as siding in settings where a rustic appearance or unusual architectural effect is desired. Side-wall shingles have a grooved surface that permits easy installation and runoff of moisture. The most common installation method used is "double coursing." This method leaves approximately ½ inch of the shingle exposed. By allowing this minimum exposure, less-expensive shingles can be used while providing maximum protection.

Side-wall shingles also come in panels for ease of installation. The panels are 8 feet long and are fastened to the structure with nails or screws. The panels are available in a wide variety of textures and finishes.

SOFFITS

A soffit panel is the part of the cornice that fits between the side wall of a structure and the fascia board. Soffits are made from ⅜ inch (A.C.) plywood, laminated wood-fiber panels, or metal. The plywood panels are cut to size and fit from standard 4 × 8-foot sheets. The laminated wood-fiber panels come in a variety of standard widths and in lengths up to 12 feet. The panels come with or without screened vents. The metal soffits come in panels that are of different standard lengths and are 9 inches wide. The panels fit into a wall hanger strip that has been nailed to the side wall and is level with the bottom edge of the fascia board. The panel is nailed into place on the outer edge through prepunched holes. The panels have the screened vents already punched in, so all that is required is to place and secure the panels and enclose the ends with a fascia cover.

WALL FLASHING AND SHEATHING

Flashing is installed over the drip caps of windows and doors. The flashing comes in various standard widths to fit the different drip caps or moldings. The flashing normally comes in 10-foot lengths and is cut to fit each window and door. It is then nailed in place using 4d rust-resistant nails. (See Figure 4–88.)

Sheathing consists of solid wood, plywood, rigid polystyrene foam, or exterior fiberboard, commonly called blackjack. It is applied to add rigidity, strength, and some insulation. Modern construction almost exclusively uses panels or sheets, since they can be installed quickly. The sheets are normally ½ inch thick and

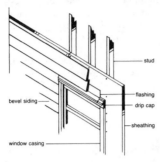

(Figure 4-88) Flashing

4 feet wide and come in lengths up to 14 feet. You should select a length that will keep any trim or cutting work to a minimum.

Plywood sheets should be installed on each side of each corner. These sheets should have a $\frac{1}{16}$-inch space between panel edges, and a $\frac{1}{8}$-inch space between panel ends. Plywood sheets provide lateral strength and eliminate the need for diagonal bracing. The blackjack (fiberboard) is installed along the remaining open areas of the walls. To nail the sheathing, use 1-$\frac{1}{2}$-inch roofing nails for $\frac{1}{2}$-inch sheathing. Provide a $\frac{1}{8}$-inch space between all edge joints when applying fiberboard sheathing.

Older methods of construction made use of wood boards or shiplap. The boards were $\frac{3}{4}$-inch thick and between 8 and 12 inches wide. The boards were applied diagonally on the studs. This is a construction process that is not commonly used except in remodeling work when it is easier to use shiplap in a small area rather than tear out a whole wall.

Another form of sheathing that is used frequently is rigid plastic foam made from polystyrene or polyurethane. This sheathing has a *R*-value of 5.50 and is used for its insulating qualities. It comes in the same sizes as the other sheathing materials. It is fragile and must be handled with care to prevent breaking or denting the surface. The foam can come with or without an aluminum backing. When a brick veneer is being applied, the aluminum-backed sheathing is used.

WINDOWS

Windows are covered openings in walls of a building that provide any or all of the following: natural light, natural ventilation, visibility, protection from the elements, and sound insulation. To be satisfactory, windows must be durable, weathertight, reasonable in cost, readily installed, and attractive. Some windows are used to a limited extent in partitions for visibility between rooms, but this section deals only with exterior windows. Windows are made from wood, steel, aluminum, stainless steel, plastic,

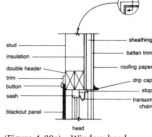

(Figure 4-89a) Window head

and bronze. They are available in a great variety of types to suit many different requirements and individual preferences.

Windows, whatever type, consist essentially of two parts, the frame and the sash. The frame is made up of four basic parts: the head, the jambs (two), and the sill. (See Figure 4–89 a–c.) The sash is the framework that holds the glass in the window. There are many different materials and variations in the

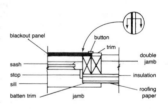

(Figure 4-89b) Window jamb

frame and sash components, but whatever the style and composition, the frame and sash serve the same purposes for every window.

HEIGHT

Window height is very important, because one of the major functions of a window is to provide a view of the outdoors. In residential construction, the standard height from the bottom side of the window head to the finished floor is 6 feet 8 inches. When this dimension is used, the heights of window and door openings will be the same. Inside and outside trim must align; ½ to ¾ inch must be added to this height for thresholds and door clearances. Window manufacturers usually provide exact dimensions for their standard units. The architectural plans will normally specify whether

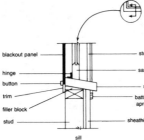

blackout panel

hinge

button

trim

filler block

stud

stop

sash

sill

batten
apron

sheathing

sill

(Figure 4-89c)   Window sill

there are to be any exceptions to the standard height. For example, if the home will be used by a disabled person who must use a wheelchair, the window heights would be lower to accommodate the view required for that person.

SIZES

Window sizes are based on four units, and you need to know how each is used. The units are the glass size, sash size, rough frame opening, and masonry or unit opening. The architectural drawings normally include a table called a window and door schedule. This schedule includes the manufacturer's numbers and rough opening sizes for each unit or combination. An identifying letter is located at each opening on the plan, and a corresponding letter is then used in the schedule to specify the required window unit and the necessary information. When studying the plans, keep in mind that the horizontal dimension is always listed first, and the height dimension is listed second for the rough opening, sash, and glass.

If the plans do not give the window and door schedule, you can secure manufacturers catalogs and study their units to select the windows needed as well as the sizes of the required rough openings.

Basic Procedures for Installing Windows

1. *Check that the rough opening is plumb, level, and the correct size.*

2. *Read the manufacturer's instructions carefully. Different manufacturers have different techniques for setting in windows.*

3. *Unpack and check the windows for any shipping damage. Leave the diagonal braces or spacer strips on the windows until installation has been completed, since they help to keep the windows square.*

4. *Check for clearance around the rough opening. You will need approximately ½-inch clearance on each side and ¾-inch clearance*

above the head. This clearance will allow you to plumb and level the window.

5. *Most windows come from the factory primed; if not, the unit can be primed at this time.*

6. *Windows are generally set in place from the outside. If the windows have been stored inside the structure, they can be easily angled through the rough opening. Set the bottom in the rough opening and then tip the top of the window in place.*

7. *Windows generally come with the outside casing attached. If the casing is not attached, it should be put in place as soon as the window has been set, leveled, and plumbed.*

8. *Insert wedge blocks under the sill and raise the frame to the correct height. Adjust the blocks until the frame is perfectly level.*

9. *Use a level to plumb the side jambs, and check that the corners are square. The sashes should be left closed and locked during installation.*

10. *Drive nails temporarily into the top of the side casing. Check the entire window once more to be sure that it is square and plumb. Open and close the sash several times to be sure that it operates smoothly. If the head sags or the jambs bow, they can be straightened with a spacer strip.*

11. *Nail the window permanently in place with aluminum or galvanized casing nails. Space the nails about 16 O.C. and be certain that they are long enough to penetrate well into the building frame. The casing nails should be driven into the casing using a nailset for the last few strokes, since the casing wood is soft and dents easily.*

12. *When installing fixed or nonmovable sash units, the installation process is generally the same as for the movable sash units. A factor that must be considered when installing fixed units is the weight of the unit. Most fixed units are made with ¼-inch insulated or thermopane glass, and this adds extra weight as compared to a regular sash unit.*

PART THREE

CARPENTRY FUNDAMENTALS

5
Units of Measure and Basic Mathematics

A solid mathematical background is required to compute the necessary measurements for a job. This includes having an aptitude for basic mathematical operations, such as addition, subtraction, division, and multiplication, as well as for the identification and use of fractions and certain formulas, such as those used in calculating surface area. Every carpenter must be able to take measurements and then make the necessary computations based on them. This chapter outlines common units of measure as well as general mathematical principles and how they can be applied.

STANDARD WEIGHTS AND MEASURES

To bring uniformity to the measuring process, the U.S. Constitution empowers Congress to set standards of weights and measures. For example, in 1866, an act of Congress set the legal equivalent of the meter at 39.37 inches.

Some of the units of measure that are considered part of the U.S. system of weights and measures and their equivalents that a carpenter would need are listed below.

Weight

16 ounces (oz)	1 pound (lb)
2,000 pounds	1 short ton

Liquid Measure

2 pints (pt)	1 quart (qt)
4 quarts	1 gallon (gal)

Length

12 inches (in)	1 foot (ft)
3 feet	1 yard (yd)

Square Measure

144 square inches (sq in)	1 square foot (sq ft)
9 square feet	1 square yard (sq yd)

Cubic Measure

1,728 cubic inches (cu in)	1 cubic foot (cu ft)
27 cubic feet	1 cubic yard (cu yd)

METRIC MEASURES

In 1975, the Metric Conversion Act was signed into law, which committed the United States to a conversion of all measurements to the metric system. The intent of the law was to bring the U.S. to the same standard as most of the industrial nations of the world within ten years. Because compliance with the law was voluntary, the conversion has progressed at a much slower rate. Since the building trades, unlike manufacturing industries, are not concerned with exporting their product, they have not found it necessary to quickly convert to metric measure.

As a carpenter, you may not work with metric measurements often; however, you should know that the metric system is a decimal (base ten) system of measurement. The standard for length is the meter, a unit that is slightly longer than a yard. While there are several subunits, you would mostly need only one, millimeters (1,000 millimeters equal 1 meter). Architectural drawings will show only those two units of length. They may be squared for area and cubed for volume, if needed.

MEASURING LENGTH AND WORKING WITH FRACTIONS

Measurement can be broadly classified as either exact or approximate. You use exact measurement when you count objects. For example, if you count 24 sheets of plywood, you know the exact number is 24 for the count. You calculate approximate measurement by using measuring instruments. For example, if you measure a piece of lumber to be 3 feet, this is only an approximate measure of its length, since all measuring instruments give only approximate results. Also, the person using the instrument cannot be perfectly accurate.

The devices you use to measure—yardstick, wooden ruler, steel rule, steel tape, framing square, etc.—have different degrees of accuracy, and the smaller the units on the measuring instrument, the more precise the measurements. For example, a carpenter's rule measures to ⅛ inch and a steel square and steel tape measure to ¹⁄₁₆ inch.

If you can read a ruler accurately, then the process of measuring will be greatly simplified. You will have to become very familiar with the various markings on the ruler or other measuring devices. You will generally be measuring to ¹⁄₁₆ inch, but in some cases you will have to measure to ¹⁄₃₂ inch.

When measuring, you should always express the measurement in its lowest terms. Thus, if you have a ⅝ inch measurement, you should convert it to ¾ inch. This is done by dividing both the numerator (the term above the line) and the denominator (the term below it) by their highest common factor (a number that can be divided evenly into both). In this case, the highest factor is 2:

$$\frac{6 \div 2}{8 \div 2} = \frac{3}{4}$$

Note that the value of a fraction is not changed when each term of the fraction is multiplied or divided by the same number.

ADDITION OF FRACTIONS AND MIXED NUMBERS

A whole number is an integer such as 0, 1, 2, 3, etc. It is a number that is not a fraction and does not contain a fraction. Mixed numbers are combinations of whole numbers and fractions, such as 1½, 2¼, 125⅜, etc. To add a series of numbers that includes mixed numbers, the first step is to add the fractions. Note that to add fractions, you must find the common denominator. For example, to add ¼ and ½, you must change the fraction ½ to ¼ by multiplying both its numerator and denominator by 2:

$$\frac{1}{2} \times \frac{2}{2} = \frac{2}{4}$$

■ **EXAMPLE**

To add the numbers 3¼, 6, 5½, 6, and 3¼, first find the common denominator for the fractions and then add the fractions:

$$\frac{1}{4} + \frac{2}{4} + \frac{1}{4} = \frac{4}{4} = 1$$

Then add the whole numbers:

$$3 + 6 + 5 + 6 + 3 = 23$$

The total sum is 24 (1 + 23).

In many problems of addition and subtraction, it will be necessary to raise fractions to higher terms, as you have done by raising ½ to ¼.

Recall that the value of a fraction is not changed when you multiply both the numerator and the denominator by the same number. Thus, if you want to change ½ to sixteenths, divide 16 by 2 and multiply both the numerator and denominator of ½ by 8:

$$\frac{1}{2} \times \frac{8}{8} = \frac{8}{16}$$

■ **EXAMPLE**

Change ½ to eighths.

$$Step\ 1 \quad 8 \div 2 = 4$$
$$Step\ 2 \quad \frac{1}{2} \times \frac{4}{4} = \frac{4}{8}$$

SUBTRACTION OF FRACTIONS AND MIXED NUMBERS

Suppose you want to subtract ½ from 3¾. The first step is to find the common denominator for the fractions:

$$\frac{3}{4} = \frac{3}{4}$$

$$\frac{2}{2} \times \frac{1}{2} = \frac{2}{4}$$

Then subtract: $3\frac{3}{4} - \frac{2}{4} = 3\frac{1}{4}$

Suppose you want to subtract ½ from 3¼. The first step of the process is again to find the common denominator for the fractions, as in the above example:

$$\frac{1}{4} = \frac{1}{4}$$

$$\frac{2}{2} \times \frac{1}{2} = \frac{2}{4}$$

However, since ²⁄₄ cannot be subtracted from ¹⁄₄, an additional step is necessary. Subtract 1 from the whole number, change it to a fraction of the common denominator, and add it to the fraction of the mixed number. Change 3¼ to 2⁵⁄₄ ($2 + \frac{4}{4} + \frac{1}{4}$) and then subtract:

$$3\frac{1}{4} = 2\frac{5}{4}$$
$$-\ \frac{1}{2} = \ \ \frac{2}{4}$$
$$\overline{\qquad\quad 2\frac{3}{4}}$$

MULTIPLICATION OF FRACTIONS AND MIXED NUMBERS

Multiplication is a short cut for adding quantities of the same value. Suppose you need to find the length of a piece of wood that has a series of eight punched holes that are ⁵⁄₈ inch apart. To find the length, multiply 8 (or its equivalent, ⁸⁄₁) by ⁵⁄₈:

$$\frac{8}{1} \times \frac{5}{8} = \frac{40}{8} = 5 \text{ inches}$$

This calculation can be simplified by means of cancellation. Cancellation is actually division of like numbers. Thus, the problem just described becomes:

$$\frac{8}{1} \times \frac{5}{8} = \frac{1}{1} \times \frac{5}{1} = 5 \text{ inches}$$

DIVISION OF FRACTIONS AND MIXED NUMBERS

When a 12-inch line is divided into 4-inch spaces, there will be three equal spaces. To divide 12 by 4, *multiply* 12 by the *reciprocal* of 4 ($\frac{1}{4}$):

$$\frac{12}{1} \div \frac{4}{1} = \frac{12}{1} \times \frac{1}{4} = \frac{12}{4} = 3 \text{ equal spaces}$$

When a 4-inch line is divided into $\frac{1}{2}$-inch spaces, the result will be eight equal spaces: To divide 4 by $\frac{1}{2}$, *multiply* 4 by the *reciprocal* of $\frac{1}{2}$ ($\frac{2}{1}$):

$$\frac{4}{1} \div \frac{1}{2} = \frac{4}{1} \times \frac{2}{1} = 8 \text{ equal spaces}$$

The reciprocal of a number is obtained by inverting the number. If the number is a whole number or mixed number, it must first be put into fractional form before the reciprocal can be obtained.

Number	Fractional form	Reciprocal
2	$\frac{2}{1}$	$\frac{1}{2}$
$1\frac{1}{2}$	$\frac{3}{2}$	$\frac{2}{3}$
$1\frac{1}{4}$	$\frac{5}{4}$	$\frac{4}{5}$

IMPROPER FRACTIONS

In some formulas you will need to substitute a mixed number, but such a number in a formula is difficult to work with. To solve the formula with ease, you need to convert your mixed number to an improper fraction. An improper fraction is one that has a numerator larger than the denominator. To make an improper fraction of a mixed number, multiply the whole number by the denominator and add it to the numerator.

Put this new numerator over the old denominator to make the improper fraction. For example, to change 2¼ to an improper fraction, multiply

$$2 \times 4 = 8$$

and add the answer to the numerator

$$8 + 1 = 9$$

to form the improper fraction

$$\frac{9}{4}$$

MIXED UNITS OF MEASURE

Most of your work will be measured in inches, feet, and yards. These units may be expressed in terms of each other, such as:

$$1 \text{ inch} = \frac{1}{12} \text{ foot} = \frac{1}{36} \text{ yard}$$
$$1 \text{ foot} = \frac{1}{3} \text{ yard}$$

Frequently, it is necessary to solve problems involving mixed units, such as when adding lengths that are expressed in feet and inches, as in 3 feet 2 inches. In performing various mathematical operations with mixed units, you must be careful to keep the units separate.

ADDITION

Although calculators are now available that add mixed units, it is a good idea to understand the mathematics involved in this type of problem solving. Suppose you need to find the length of a wall on the outside of a house. The measurements are as follows: 6 feet 7 inches, 14 feet 5 inches, 8 feet 9 inches, and 9 feet 10 inches.

Add the units separately to arrive at 37 feet 31 inches. Now convert the 31 inches to feet and inches by dividing 31 by 12:

$$\frac{31}{12} = 2\frac{7}{12} = 2 \text{ feet } 7 \text{ inches}$$

The 2 feet 7 inches can now be added to the number of feet already obtained: 2 feet 7 inches + 37 feet = 39 feet 7 inches.

SUBTRACTION

If a length of insulation that is 39 feet 7 inches is cut from a 250 foot roll, what length will remain? To find the length remaining, subtract inches from inches and feet from feet. Since 7 inches cannot be subtracted from 0 inches, it is necessary to change the 250 feet to 249 feet 12 inches.

$$
\begin{array}{r}
249 \text{ feet } 12 \text{ inches} \\
- \quad 39 \text{ feet } 7 \text{ inches} \\
\hline
210 \text{ feet } 5 \text{ inches}
\end{array}
$$

The amount of insulation left on the roll is 210 feet and 5 inches.

MULTIPLICATION

Suppose you have to calculate the total length of thirteen rods that are each 2 feet 6 inches long. It is easiest to multiply the units separately by 13 and then convert inches to feet:

$$
\begin{array}{r}
2 \text{ feet } 6 \text{ inches} \\
\times \quad\quad 13 \\
\hline
26 \text{ feet } 78 \text{ inches}
\end{array}
$$

Change inches to feet:

$$\frac{78}{12} = 6\frac{6}{12} = 6 \text{ feet } 6 \text{ inches}$$

Add 26 feet to 6 feet 6 inches to arrive at the total length of 32 feet 6 inches.

In addition to multiplying mixed units, you will often have to determine the square or cube of a number. Squares and cubes are needed when using formulas. The square of a number is that number multiplied by itself. For example the square of 25 is

$$25 \times 25 = 625$$

The cube of a number is the square of the number multiplied by the number. For example, the cube of 8 is

$$8 \times 8 = 64$$
$$64 \times 8 = 512$$

or

$$8 \times 8 \times 8 = 512$$

DIVISION

In dividing mixed units, it is easiest to change the mixed units to numbers containing the lowest units before dividing.

■ **EXAMPLE**

A board 15 feet 8 inches is to be divided into eight equal parts. Find the length of each part and express the answer in feet and inches:

Step 1 15 feet 8 inches = (15 × 12) + 8 = 188 inches
Step 2 188 ÷ 8 = 23⅛ = 23½ inches
Step 3 Convert 23½ inches into feet and inches

Each of the eight equal parts of a board 15 feet 8 inches is 1 foot 11½ inches long.

6
Identification of Woods and Lumber Terminology

Since the beginning of time, wood has been used to build various types of structures. It was first used because it was one of the easiest materials to work with, it was available in large quantities, and it was a renewable resource. Early forms of lumber consisted of trimmed and slabbed logs.

Wood is still the most commonly used construction material, and generally for the same reasons. Today, however, different forms of wood are used in construction, ranging from sheets of processed wood chips to traditional dimension lumber.

LUMBER CHARACTERISTICS

Over 600 varieties of trees grow in the United States, and hundreds of others are found throughout the world. Only a few varieties, however, are suitable for construction purposes, and even fewer are available in sufficient quantities to be of commercial importance. The species that are commonly used in construction do not have identical characteristics, so it's important to know the prominent features of different wood types in order to make the appropriate lumber selection for specific construction needs.

HARDWOODS AND SOFTWOODS
Woods are generally divided into two broad categories: hardwoods and softwoods. Hardwoods are obtained from deciduous trees, which have broad leaves that are generally shed at the end of each growing season. Hardwoods are used for flooring, wall and ceiling paneling, decorative moldings, and furniture.

Softwoods come from coniferous trees, or evergreens. These trees have needlelike leaves that are generally not shed at the end of each growing season. Softwood is invariably used in structural applications that include the framing, sheathing, and blocking of buildings. Softwood is also used for the same purposes as hardwood, such as paneling and molding, but softwood does not offer the choice of colors or quality of hardwood.

In classifying woods as hardwood or softwood, there is no relationship between these designations and the actual hardness or softness of the wood. Basswood, for example, is classified as a hardwood, but it is actually softer in texture than most softwoods. The classification of wood relates to the species and cell structure of the wood.

WOOD GROWTH AND STRUCTURE

Both hardwoods and softwoods are exogenous; that is, they are formed by the addition of a layer of new wood each year around a small central core called the pith. Each new layer in the cross section of the tree appears as an additional annual ring surrounding the older wood. The cross section of a tree consists of an exterior bark over annual growth rings surrounding a pith at the center that varies in diameter from 1/20 inch to nearly 1/4 inch. The annual rings near the outside form the sapwood and are lighter than those near the pith.

In the growing tree dense wood fibers form lines that radiate out from the center of the tree and extend through the growth rings. These lines are called medullary rays. In some wood, the medullary rays are quite prominent. Quarter-sawed oak, for example, is cut to expose the beauty of the medullary rays.

LUMBER CUTS

Wood is cut into dimension lumber in several different ways. The cutting method is determined by the use that will be made of the wood. If the wood is going to have the grain exposed as a decorative feature within the structure, the log is sawed in one way; if strength

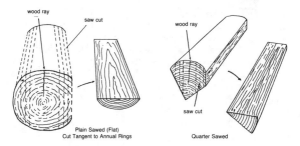

(Figure 6-1) Different cuts of wood

is the main concern, then another sawing method is used. Also, logs that are going to be used for veneer are sliced, rather than sawed, into thin sheets.

Another factor that determines the type of cutting method is cost. Flat-sawed wood is cheaper to produce, since there are fewer operations required to bring the lumber to dimension. Quarter-sawed lumber is more expensive because each log must be run through the sawing operation twice: first to quarter the log, and then to cut each quarter log to dimension. This method is commonly used when cutting hardwood logs into flooring because it shows the beauty of the grain.

FLAT-SAWED LUMBER

When wood is flat-sawed, it is cut in parallel slices that are called flat-grain lumber in softwoods and plain-sawed lumber in hardwoods. There is less waste in cutting flat-sawed lumber, and, consequently, it is cheaper than quarter-sawed lumber.

The growth-ring patterns are much more distinct in flat-sawed lumber than in quarter-sawed lumber. (See Figure 6–1.) These patterns are especially emphasized in many hardwoods. However, flat-sawed lumber has certain disadvantages. It tends to warp, check, split, and show separations of growth rings. It also shrinks and swells in width.

QUARTER-SAWED LUMBER

Quarter-sawed lumber is cut so that the annual rings form an angle of more than 45 degrees with the surface of the board. It is more expensive than flat-sawed lumber because of the additional labor involved in the cutting process. However, this lumber has several advantages. It swells and shrinks less than flat sawed lumber and develops fewer cracks and checks during seasoning and use. It is usually denser and allows less water to pass through it. The denser, more uniform texture of quarter-sawed lumber produces a surface that wears uniformly, which is a necessity for softwood flooring and decking as well as for areas that are subject to weathering.

LUMBER DIMENSIONS

Lumber is usually sawed in dimensions of even inches in thickness and width and in 2-foot increments of length, generally from 8 to 20 or 24 feet. These are nominal dimensions of the lumber. However, the actual sizes, due to sawing, planing, and surfacing, are ¼ to ¾ inch less than the nominal dimensions. Standard sizes of lumber have been established by the United States Department of Commerce for various types and species.

In 1966, a committee of the National Lumber Manufacturers recommended to the Department of Commerce that the sizes of finished softwood lumber be revised. The new sizes were incorporated into the new lumber standard PS 20-70. There have been many problems in specifying sizes of lumber in the past. Green lumber cut at mills to the standard sizes may arrive on the job site undersize, whereas boards cut after they have been seasoned, or from dry lumber, must be cut to exact dimensions. The new sizes are smaller than those previously used, but they are calculated by using a standard moisture content of 15 percent. Each size is then correlated with the specific moisture content expected when lumber is incorporated into a building.

Rough lumber is unfinished as it comes from the saw. Sizes of finished lumber are the dimensions of the lumber after it has been run through a planer, or dressed. Lumber may be surfaced on one or more sides and is designated as shown in Table 6–1. Boards may be further worked by running them through matching machines, stickers, or molders. These boards, usually used as siding or sheathing, are classed as shiplap, tongue and groove, or patterned. The boards may be square-ended or they may be end-matched. End-matched boards have tongues and grooves cut into the ends of the pieces to form the joints.

TABLE 6–1
DESIGNATIONS FOR DRESSED LUMBER

Designation	Description
S1S	Surfaced one side
S2S	Surfaced on two sides
S1E	Surfaced on one edge
S2E	Surfaced on two edges
S1S1E	Surfaced on one side, one edge
S2S1E	Surfaced on two sides, one edge
S4S	Surfaced on four sides

UNITS OF MEASURE

The board foot is the unit of measure for lumber. It is the quantity of lumber contained in a piece of rough green lumber 1 inch thick, 12 inches wide, and 1 foot long, or its equivalent (144 square inches) in thicker, wider, narrower, or longer lumber. Lumber less than 1-inch thick is considered 1 inch thick.

In referring to commercial lots of lumber, this method of measurement is called board measure (b.m.). The common unit is 1,000 board feet, designated as M. For example, a lot of lumber that contains 25,000 board feet is designated as 25 M.b.m.

Moldings are measured by the lineal foot, and shingles are measured by the number of pieces of a specified length.

LUMBER GRADES

Lumber is graded according to its strength, appearance, or usability. The United States Department of Agriculture, the National Bureau of Standards, and lumber associations have for many years attempted to simplify sizes, nomenclature, and grades of lumber and obtain uniformity of practice. (See Figure 6–2.)

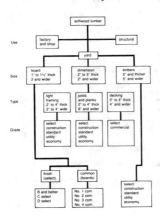

(Figure 6-2)   Grades of softwood

When lumber is cut from a log, the individual pieces vary in appearance and strength, depending on the imperfections and the defects in each piece. Grades of lumber are based on the number and placement of such defects as knots, checks, splits, and pitch pockets. The highest grades are practically free of these defects, while each successively lower grade allows more imperfections. This does not preclude the use of the lower grades for many purposes. A piece of softwood lumber with knots is not necessarily seriously defective.

HARDWOOD GRADES

Hardwood grades are based on the amount of usable lumber in each piece of a standard length, from 4 to 16 feet. The hardwood is inspected on the poorest side of the piece, and the grade is based on the appearance of this side.

Hardwoods are classified as firsts, seconds, selects, and Nos. 1, 2, 3A, and 3B common. In general each lower grade yields smaller portions of clear pieces. Firsts and seconds are nearly always combined into one grade (FAS). Hardwoods whose unique appearance is based on worm holes and other imperfections are called sound wormy and are classed as No. 1 common.

SOFTWOOD GRADES

Softwood grades are divided into three basic classifications: yard, structural, and factory and shop. Yard lumber is further classified as either selects and finish or boards. The grades for each, as established by the Western Wood Products Association (WWPA) and the West Coast Lumber Inspection Bureau (WCLIB) are shown in Table 6–2.

TABLE 6–2
SOFTWOOD LUMBER GRADES

Grade	Uses
Yard	In general, used for framing in residences and other light construction.
Structural	Pieces 2 inches or more in thickness graded according to the actual working stresses resisted by an entire piece of lumber; working stresses vary with location and number of knots, checks, and splits, direction of grain, and presence of wane
Factory and shop	Finish lumber used for sash doors, and trim; graded according to the number of smaller pieces that can be obtained from each piece of lumber

Common lumber is another classification system used in softwood grading. Grades in this system run from No. 1, which contains tight knots, few blemishes and is suitable for natural knotty finish or paint, to No. 5, the lowest grade, used where strength and appearance are not essential.

Most lumber is graded by the "Grading Rules for Western Lumber," as set forth by the Western Wood Products Association. The WWPA is an association of lumber manufacturers who work to make lumber grading and standards consistent. A WWPA grading stamp contains five basic elements of information about the lumber

a. WWPA certification mark. Certifies Association quality supervision. (WWP) is a registered trademark.

b. Mill identification. Firm name, brand or assigned mill number. A list of mills, by number is available from WWPA offices.

c. Grade designation. Grade name, number or abbreviation.

d. Species identification. Indicates species by individual species or species combination.

e. Condition of seasoning. Indicates condition of seasoning at time of surfacing.
S-DRY — 19% maximum moisture content
MC-15 — 15% maximum moisture content
S-GRN — over 19% moisture content (unseasoned)

Inspection Certificate
When an inspection certificate is issued by the Western Wood Products Association is required on a shipment of lumber and specific grade marks are not used, the stock is identified by an imprint of the Association mark and the number of the shipping mill or inspector.

Grade Stamp Facsimiles
WWPA uses a set of marks similar to the randomly selected examples shown on the reverse side, to identify lumber graded under its supervision.

(Figure 6-3)   Grade marks

that has been graded and shipped. (See Figure 6–3.) These elements are:

1. WWPA certification mark (a),
2. mill identification (b),
3. grade designation (c),
4. species identification (d), and
5. condition of seasoning (e).

LUMBER DEFECTS

Lumber defects can occur naturally as a result of insects, injury or disease to the tree. Other defects result from improper seasoning or the milling process.

Check. This is a lengthwise separation of the grain, usually occurring through the growth rings as a result of seasoning.

Decay. Decay of wood is caused by fungi that feed on the cell walls. It is also called dote or rot. Decayed wood is often sought by woodturners and furniture makers because of the unique patterns created by the fungi.

Hit and miss. This is a series of skips occurring over surfaced areas.

Holes. These are openings in the lumber that are caused by boring insects or worms.

Knots. Knots result from a branch or limb becoming embedded in the tree during the growth process. When the knots appear in sawed lumber, they may show up as decayed knots (open areas in the lumber), tight knots (those that are firm and will stay in place), or spike knots (those that are sawed in a lengthwise direction). (See Figure 6–4.)

Machine burn. This is a darkening or charring of the wood caused by the overheating of machine knives.

Machine gouge. A machine gouge is a groove that occurs when the machine cuts below the desired line of cut.

Pitch pockets. These are openings between growth rings that usually contain or have contained resin or bark or both.

Shake. A shake is a lengthwise grain separation between or through the growth rings. It may be further classified as ring shake or pith shake.

Skip. A skip is an area on a piece that was not surfaced.

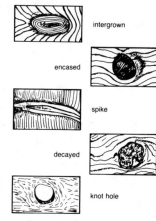

(Figure 6-4)   Knots

Split. This is a lengthwise separation of the wood extending from one surface through the piece to the opposite surface or to an adjoining surface.

Stain. This is a discoloration on or in lumber other than its natural color.

Torn grain. This is the part of the wood that is torn out in dressing.

Wane. This is bark or the lack of wood, from any cause, on the edge or corner of a piece.

Warp. A warp is any variation from a true or plane surface and includes a bow, crook, and cup in any direction. (See Figure 6–5.)

PLYWOOD

Plywood is a glued-laminated wood product that is made from thin sheets of wood that are $\frac{1}{10}$ to $\frac{1}{4}$ inch thick. Thin sheets of wood or veneers are peeled from logs that are mounted in giant lathes. The veneers are bonded together with glue under high pressure, with the grain of alternate layers running in perpendicular directions. The resulting composite material is therefore stronger than solid wood. Plywood is strong across the panel as well as along its length.

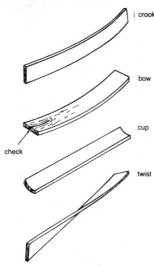

(Figure 6-6)   Warps

Plywood resists splitting, checking and splintering and is more dimensionally stable than solid wood. Plywood has less warp and twist than most solid-wood panels of comparable size. It is easy to work with ordinary tools and lends itself to speedy fabrication.

Plywood is generally produced in 4-by-8-foot sheets and standard thickness of $\frac{1}{8}$, $\frac{1}{4}$, $\frac{5}{16}$, $\frac{1}{2}$, $\frac{5}{8}$, $\frac{3}{4}$, 1, and $1\frac{1}{8}$ inches. Plywood may be custom ordered in lengths up to 50 feet if needed. It may be three-ply, five-ply, or seven-ply, and the wood sheets or plies are bonded together in odd-numbered layers, so that the outside layers on both sides will have the grain running with the length of the sheet. This minimizes the warpage that could occur if the outer layers were at right angles. Imported plywoods, such as Baltic birch, come in nine plies per $\frac{1}{2}$-inch sheet and up to thirteen plies in $\frac{3}{4}$-inch thick sheets.

HARDWOOD PLYWOOD

The term hardwood plywood denotes a wide range of products, from plywood in which all plies are hardwood veneer to plywood with hardwood outer layers and a core of softwood veneer, lumber, or particleboard.

Many expensive and rare hardwoods have color, figure, or grain characteristics that make them highly prized for paneling and cabinetwork. Modern methods of peeling or slicing thin veneers from rare and exotic woods have made it possible to use these woods in moderately priced construction. Imported hardwood accounts for over 60 percent of the American market.

Hardwood plywood is graded using a system developed by the Hardwood Plywood Institute for backs and faces. A grading specification of 1 to 2 indicates a good face with grain carefully matched and a good back but without careful grain matching. A No. 3 back permits noticeable defects and patching but is generally sound. A special or premium grade of hardwood is known as architectural or sequence-matched hardwood plywood. This is usually a special order because it requires careful layout at the mill. Hardwood plywood is most often used on the job site to custom-build bookcases and cabinets and in workshops to build furniture for homes and offices.

SOFTWOOD PLYWOOD

Most softwood plywood is constructed of Douglas fir, although southern pine, western larch, western hemlock, white fir, cedar, and other species are also used. There are two basic types of plywood: exterior and interior. Exterior plywood, which is bonded together with waterproof glues, is used for siding, concrete forms, and other construction where it will be exposed to the weather or excessive moisture. Interior plywood is made with glues that are not waterproof and is used for cabinets and other inside construction where the moisture content of the panels will not exceed 20 percent.

Softwood plywood is manufactured in accordance with United States Product Standards as established by the U.S. Department of Commerce and the plywood manufacturers. This standard provides a system for designating the species, strength, type of glue, and appearance. Plywood is marked under this system to let you know

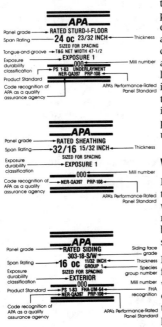

(Figure 6-6) APA registered trademarks

the grade, type, and quality of plywood you are using. On each sheet of plywood, there is a stamp that has a number of codes on it that tell you all about that sheet. There is a letter stamped on the sheet that is the grade for the quality of the sheet, and a name indicating the panel's intended use or performance rating. (See Figure 6–6.)

VENEER GRADES

Letters are used to represent the veneer grades of the plywood, and these letter designations are described as follows by the American Plywood Association.

N. Smooth surface "natural finish" veneer; select, all heartwood or all sapwood, free of open defects; allows not more than six repairs, wood only, per 4-by-8-foot panel; made parallel to grain and well matched for grain and color.

A. Smooth, paintable; not more than eighteen neatly made repairs, boat, sled, or router type, and parallel to grain, permitted; may be used for natural finish in less demanding applications.

B. Solid surface; shims, circular repair plugs, and tight knots to 1 inch across grain permitted; some minor splits permitted.

C (plugged). Improved C veneer with splits limited to ⅛ inch width and knotholes and borer holes limited to ½ by ½ inch; admits some broken grain; synthetic repairs permitted.

(Figure 6-7)   Grade marks

C. Tight knots to 1½ inch; knotholes to 1 inch across grain and some to 1½ inch of total width of knots and knotholes is within specified limits; synthetic or wood repairs; discoloration and sanding defects that do not impair strength permitted; limited splits allowed; stitching permitted.

D. Knots and knotholes to 2½-inch width across grain and ½ inch larger within specified limits; limited splits allowed; stitching permitted; limited to interior use; exposure 1 and exposure 2 panels.

APA A-A

TYPICAL TRADEMARK

A·A · G-1 · EXPOSURE 1·APA · 000 · PS1-83

Use where appearance of both sides is important for interior applications such as built-ins, cabinets, furniture, partitions; and exterior applications such as fences, signs, boats, shipping containers, tanks, ducts, etc. Smooth surfaces suitable for painting. EXPOSURE DURABILITY CLASSIFICATIONS: Interior, Exposure 1, Exterior. COMMON THICKNESSES: 1/4, 11/32, 3/8, 15/32, 1/2, 19/32, 5/8, 23/32, 3/4.

APA A-B

TYPICAL TRADEMARK

A·B·G-1·EXPOSURE1·APA · 000 · PS1-83

For use where appearance of one side is less important but where two solid surfaces are necessary. EXPOSURE DURABILITY CLASSIFICATIONS: Interior, Exposure 1, Exterior. COMMON THICKNESSES: 1/4, 11/32, 3/8, 15/32, 1/2, 19/32, 5/8, 23/32, 3/4.

APA A-C

TYPICAL TRADEMARK

APA

A-C GROUP 1

EXTERIOR

000
PS1-83

For use where appearance of only one side is important in exterior applications, such as soffits, fences, structural uses, boxcar and truck linings, farm buildings, tanks, trays, commercial refrigerators, etc. EXPOSURE DURABILITY CLASSIFICATION: Exterior. COMMON THICKNESSES: 1/4, 11/32, 3/8, 15/32, 1/2, 19/32, 5/8, 23/32, 3/4.

APA A-D

TYPICAL TRADEMARK

APA

A-D GROUP 1

EXPOSURE 1

000
PS1-83

For use where appearance of only one side is important in interior applications, such as paneling, built-ins, shelving, partitions, flow racks, etc. EXPOSURE DURABILITY CLASSIFICATION: Interior, Exposure 1. COMMON THICKNESSES: 1/4, 11/32, 3/8, 15/32, 1/2, 19/32, 5/8, 23/32, 3/4.

APA B-B

TYPICAL TRADEMARK

B·B·G-2·EXPOSURE1·APA · 000 · PS1-83

Utility panels with two solid sides. EXPOSURE DURABILITY CLASSIFICATION: Interior, Exposure 1, Exterior. COMMON THICKNESSES: 1/4, 11/32, 3/8, 15/32, 1/2, 19/32, 5/8, 23/32, 3/4.

APA B-C

TYPICAL TRADEMARK

APA

B-C GROUP 1

EXTERIOR

000
PS1-83

Utility panel for farm service and work buildings, boxcar and truck linings, containers, tanks, agricultural equipment, as a base for exterior coatings and other exterior uses or applications subject to high or continuous moisture. EXPOSURE DURABILITY CLASSIFICATION: Exterior. COMMON THICKNESSES: 1/4, 11/32, 3/8, 15/32, 1/2, 19/32, 5/8, 23/32, 3/4.

APA B-D

TYPICAL TRADEMARK

APA

B-D GROUP 2

EXPOSURE 1

000
PS1-83

Utility panel for backing, sides of built-ins, industry shelving, slip sheets, separator boards, bins and other interior or protected applications. EXPOSURE DURABILITY CLASSIFICATION: Interior, Exposure 1. COMMON THICKNESSES: 1/4, 11/32, 3/8, 15/32, 1/2, 19/32, 5/8, 23/32, 3/4.

APA UNDERLAYMENT

TYPICAL TRADEMARK

APA

UNDERLAYMENT

GROUP 1

EXPOSURE 1

000
PS1-83

For application over structural subfloor. Provides smooth surface for application of carpet and pad and possesses high concentrated and impact load resistance. EXPOSURE DURABILITY CLASSIFICATION: Interior, Exposure 1. COMMON THICKNESSES[4]: 11/32, 3/8, 1/2, 19/32, 5/8, 23/32, 3/4.

APA C-C PLUGGED

TYPICAL TRADEMARK

APA

C-C PLUGGED

GROUP 2

EXTERIOR

000
PS1-83

For use as an underlayment over structural subfloor, refrigerated or controlled atmosphere storage rooms, pallet fruit bins, tanks, boxcar and truck floors and linings, open soffits, and other similar applications where continuous or severe moisture may be present. Provides smooth surface for application of carpet and possesses high concentrated and impact load resistance. EXPOSURE DURABILITY CLASSIFICATION: Exterior. COMMON THICKNESSES[4]: 3/8, 1/2, 19/32, 5/8, 23/32, 3/4.

APA C-D PLUGGED

TYPICAL TRADEMARK

APA

C-D PLUGGED

GROUP 2

EXPOSURE 1

000
PS1-83

For built-ins, cable reels, walkways, separator boards and other interior or protected applications. Not a substitute for Underlayment or APA Rated Sturd-I-Floor as it lacks their puncture resistance. EXPOSURE DURABILITY CLASSIFICATION: Interior, Exposure 1. COMMON THICKNESSES: 1/4, 11/32, 3/8, 1/2, 19/32, 5/8, 23/32, 3/4.

(1) Specific plywood grades, thicknesses and exposure durability classifications may be in limited supply in some areas. Check with your supplier before specifying.

(2) Sanded exterior plywood panels. C-C Plugged, C-D Plugged and Underlayment grades can also be manufactured in Structural I (all plies limited to Group 1 species) and Structural II (all plies limited to Group 1, 2 or 3 species).

(3) Some manufacturers also produce plywood panels with premium N-grade veneer on one or both faces. Available only by special order. Check with the manufacturer.

(4) Panels 1/2 inch and thicker are span rated and do not contain species group number in trademark.

(Figure 6-8)   Sanded and Touch sanded plywood

RECONSTITUTED WOOD-BASED PANELS

With the development of better glues and manufacturing processes, the manufacturers of reconstituted wood-based panels have been able to deliver a quality product that is less expensive, as a result of using trees that were not commercially usable before, and that is competitive with plywood. Since 1973, the production of particleboard has increased ten times, and it is expected that by 1990 "waferboard" production will exceed Douglas fir plywood. These panel products have the advantage of being manufactured under controlled conditions that can create panels that are stronger, more versatile, less subject to warpage, and even more weather resistant than plywood. In addition, the manufacturers can introduce into

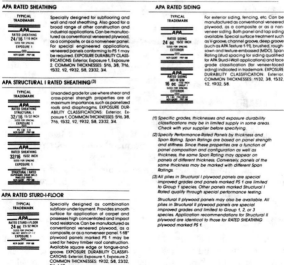

(Figure 6-9)   APA performance rated panels

the resins, during manufacturing, agents that resist moisture, fungus, and insects and that are fire retardant. There are six types of reconstituted wood-based panels recognized by the American Plywood Association with which you should be familiar. These panels normally come in 4-by-8-foot sheets, though they can be ordered in various lengths. The thickness of the panels range from ¼ to 1¼ inches.

COMPOSITE PANELS

Panels made by combining a reconstituted wood core with veneer faces are classed as composite panels. Waferboard or flakeboard usually is used as the core. However, any type of particleboard may be used. These panels are exceptionally warp-free and have a dimensional stability that make them valuable for sliding and hinged cabinet doors.

FIBERBOARD

The wood is broken down into individual fibers and combined with synthetic resins in such a way that they are linked together in a natural woodlike bond. Medium density fiberboard (MDF) is used widely as a core material in plywood composite panels used for furniture.

FLAKEBOARD

Flakeboard is similar to waferboard except that the particles are smaller and cut with the grain of the wood. The flakes are thin and long and can be aligned to produce "oriented strand board" (OSB). The oriented flakes are placed in perpendicular layers similar to plywood. The distinction between waferboard and flakeboard may be very slight. The outside layers may consist of larger wafers and appear to be waferboards.

OXBOARD

Oxboard is an oriented strand board (OSB) that has five layers and is bonded with liquid phenolic resin. It is comparable in strength to plywood and is being used where sheathing plywood has been traditionally used.

PARTICLEBOARD

This term covers a large number of products that are made up of smaller wood particles bonded with urea-formaldehyde. A product designated as structural particleboard has been developed that uses phenol-formaldehyde or other nonurea resins in its formulation.

WAFERBOARD

Waferboard is made up of large wood flakes. These flakes are generally 1¼ inch or longer and are either square or oblong. The thickness of the flakes is usually ¹⁄₂₀ to ¹⁄₄₀ of length. The particles are laid out randomly, which produces a board with strength that is equal in all directions and is not dependent on the direction of the grain.

SEASONING

The process of drying lumber to the point at which it is ready for use is called seasoning. When trees have been cut into lumber, the green lumber is stored at the mill or in lumber yards to air-dry. In order for lumber to dry properly, it must be stacked in covered piles, with each successive layer separated by 1-inch strips so that air can flow between the layers. The time necessary to bring lumber to the proper moisture content varies with the area and the weather conditions at the place of storage. In the United States, lumber is seldom stacked for more than three or four months.

Wood will absorb moisture after it has dried, causing it to swell. The best practice is to dry structural lumber until it reaches a moisture content equal to the average moisture content of air in the area where it will be used. The recommended moisture content varies from 7 percent in dry southwestern states to 18 percent in damp coastal portions of the country. It is a good practice to store lumber on the job site for several months before it is incorporated into a building. This will allow the lumber to reach an equilibrium so that the shrinking and swelling will be minimized.

KILN DRYING

After a period of seasoning, lumber can be brought to the desired moisture level by kiln drying. Kiln drying is accomplished by placing the lumber in an oven, or kiln, and exposing it to elevated temperatures of 70 to 120 degrees Farenheit for a period of between four and ten days. The temperature has to be carefully controlled, or the lumber may split, check, or warp.

WOOD PRESERVATIVES

Wood and wood products must be protected from weather, insects, and fungi if they are going to last for any period of time. Preservatives used to combat the destructive agents of wood can be generally classified as coal-tar-based products such as creosote, water-soluble metallic oxides or salts such as zinc chloride and copper salts, and solvent-soluble products such as chlorinated phenols.

Creosote forms an effective barrier against decay and insects. Creosote's odor is objectionable, so its uses are often limited to timber that is used for pilings, power poles, and waterfront structures. Creosote is insoluble in water, so it will not wash out of timbers used in structures subject to the action of water. Creosoted wood cannot be painted successfully.

Water-soluble salts are either painted on the surface or forced into the wood by pressure to give the wood protection against decay and insects. Lumber may be successfully painted after treatment with these salts.

Solvent-soluble materials such as pentachlorophenol are used as preservatives on fence posts, structural lumber, sash, and doors. When the solvent evaporates, a coating of nonsoluble material is left on the wood. The wood may be painted, but success in painting is not universal. Solvents vary from No. 1 oil (unpaintable) to mineral spirits (sometimes paintable) to methylene chloride and liquified petroleum gas (usually paintable). Much controversy has surrounded the use of this group of preservatives, particularly methylene chloride, because of the danger to the user. Product labels contain manufacturers' warnings concerning the hazards of these products, and extremely cautious use is recommended.

7
Selection of Appropriate Materials

A carpenter works with a number of different materials, besides wood and wood-based products. The most common of these materials are:

- wallboard and sheathing,
- insulation boards and blankets,
- shingles of asphalt, metal, and fiberglass,
- metal flashing materials,
- caulking materials, and
- resilient flooring materials and carpeting.

These materials, their characteristics and purposes, were discussed in depth when used in the construction process, so they are mentioned only in passing at this point.

There are a number of other nonwood materials with which you must be familiar in order to keep up to date with the construction industry. These materials are described in the following pages.

ADHESIVES

Adhesives come in many different forms and are designed for many different kinds of application. You must carefully consider how the article or area to which you are applying adhesive will be used, as well as whether the gluing process will involve clamping or applying pressure and whether temperature will affect the adhesive application.

To achieve satisfactory results from any adhesive, follow the manufacturer's instructions precisely, making sure joints are clean, dry, and snug-fitting. The following list outlines some of the uses and types of adhesives that are commonly used in construction.

TYPES AND USES

CONTACT CEMENT

This type of adhesive is used when a strong bond is needed, as in bonding laminates to countertops. The adhesive is applied to both surfaces and allowed to dry until neither surface is sticky to the touch. The two surfaces are then put together. A strong permanent bond results, but in making the bond there are no second chances. The materials must be in place when pressed together. They cannot be moved once they have been placed. Contact cement is also a good choice when bonding together plastic foam (insulation around a basement wall), hardboard, or metal to wood. Excess adhesive can be removed with a special solvent, such as lacquer thinner. This adhesive is extremely flammable and must be used in a well-ventilated area.

EPOXY ADHESIVES

These are two-part adhesives (one part hardener and one part resin) that can be used to hold almost any kind of material. They create a strong bond and are moisture resistant. Also, they do not set up instantly, which is an advantage if the parts that are to be bonded must remain movable until the final fit is made. Epoxies have varying drying or setting times, ranging from five minutes to eight hours. Epoxies are too expensive for general construction use, but they work very well when a special situation calls for extra holding power.

MASTIC ADHESIVES

These heavy, pasty adhesives are applied with caulking guns or trowels. They are commonly used to glue ceiling tile, floor tile, wallboards, and wood paneling. These adhesives create a strong bond, are moisture proof, and can be worked for several minutes until the workpiece is positioned as desired. The excess glue is easily removed with a damp cloth or with standard mastic solvent.

POLYVINYL RESIN EMULSION ADHESIVES

These are the white glues that come in squeeze bottles and are used for just about everything from gluing paper together to holding wood joints. They are an excellent choice for interior construction. They dry clear and form a strong bond. However, they do not hold up under moist or damp conditions. Excess glue should be removed with a damp cloth while the glue is still wet. Since these glues do not allow stain or finish to penetrate, a clear or white spot will result if the excess glue is not removed. If the glue comes in contact with metal (such as a clamp) and wood, a chemical reaction will create a dark or black spot on the wood.

SILICONE SEALANTS

These adhesives come in tubes and are most commonly used as sealing compounds. Typical applications are made around sinks and bathtubs, where the sink top or tub meets the wall, and thresholds, where a seal is needed to keep moisture away from the structure. The excess should be removed before the sealant is allowed to harden. These sealants come in colors to match the work or clear.

UREA-FORMALDEHYDE-RESORCINOL GLUES

These glues create a strong bond when gluing wood and require clamping. The excess glue can be wiped away with a damp cloth. They may come as powders that require mixing with water or a provided catalyst. Drying time for these glues ranges from three to ten hours, and they should be used in an area where the temperature is not less than 70 degrees.

CAULKING COMPOUNDS

Caulking is made in many different compounds, and the selection of the appropriate one depends on the cost and the type of seal that is desired in the work area. Caulking is used primarily to seal out moisture and to prevent air infiltration. Caulking comes in tubes and propellant cans. The tube caulking is applied by inserting the tube in a caulking gun and applying pressure. The propellant-can caulking comes as a foam that flows into the area and then hardens. New caulking products are constantly introduced into the market, so you should check with your building supplier to make sure you are aware of the latest products available.

METAL FASTENERS

NAILS

Nails are the most commonly used metal fasteners in construction. They are available in a wide variety of shapes, sizes, and types. The choice of specific nail size, shape, and type depends on the holding power required, the material being fastened, and the surface finish desired.

Nail length is designated in inches and also by "penny" size, a term that originally related to the price per hundred but now signifies length only. Common nails, for example, are available in lengths from 1 inch or 2 penny (abbreviated 2d), to 6 inches, or 60 penny.

Except for some special-purpose nails, the diameter increases with length—a 6-inch common nail is nearly four times the diameter of a 1-inch nail. Special-purpose nails, however, may come in only one size—as flooring brads do, for example—or in several lengths but only a single diameter, as some shingle nails do, depending on their purpose.

As a general rule, nails should be driven through the thinner piece of wood into the thicker one. The nail should be three times as long as the thickness of the thin piece through which it passes; thus, two-thirds of the nail will be in the thicker piece of wood for maximum holding power. Also, certain nails work better for certain jobs. For example, the common nail has a heavy cross section and is designed for heavy framing, while the box nail is thinner and is used for toenailing in frame construction and light work.

Nails are made from different materials based on their use. The most common material is iron or steel. These nails are used where rust is not a problem. Where rust or weather is a concern, nails made from copper, aluminum, bronze, or stainless steel are used. Nailing into masonry walls requires that a special kind of nail be used. These nails are round, square, and fluted and are made from hardened and tempered steel. They are often used to fasten framing parts such as sills, furring, and window and door trim to masonry and concrete.

These nails can be driven directly into concrete-block or poured-concrete walls and floors. They are usually fired into the concrete using a power driver that contains an explosive charge and sets the nails completely into the wall with a minimum of effort. However, be sure to wear eye protection when using the power driver and use the correct power charge for the material being fastened or nailed. The power charges are color coded for ease of selection.

Nails used in power nailers come in clips that are inserted into the gun. The selection of nails is based on the job to be completed and the factors that must be considered are the same as those for nails that are hand driven.

Nailheads are a consideration in selecting the proper nails for the job. Large heads hold best because they spread the load over a wider area, resisting pullthrough. The heads of finishing nails, conversely, pull through wood quite readily. This is sometimes a welcome weakness, because it permits trim and cabinetwork to be disassembled by pulling the nailheads with a minimum of surface marring.

SCREWS

Screws are classified according to the shape of the head, surface finish, and the material from which they are made. Wood screws are used if the holding power required is greater than the holding power provided by nails or if the fastened parts may need to be taken apart at a later date. Wood screws most often are used for interior construction and cabinetwork. Their sizes are determined by length and diameter (gauge number), and they are available in lengths from ½ to 6 inches and in gauge numbers from 0 to 30. Screw lengths up to 1 inch increase by ⅛-inch increments, screws 1 to 3 inches long increase by ¼-inch increments, and screws 3 to

6 inches long increase by ½-inch increments. The gauge number can vary for a given length of screw. For example, a 1-inch screw is available in gauge numbers of 3 through 20. The lower the gauge number, the thinner the screw diameter. A 6-gauge screw is smaller in diameter than a 10-gauge screw.

The most commonly used wood screws are made of mild steel with a zinc-chromate finish. They come with flat heads (F.H.), oval heads (O.V.), or round heads (R.H.). Nickel, chromium-plated, and brass screws are available for special jobs. These screws also are available in different head shapes and sizes. Wood screws are usually sold in box lots of 100 screws per box.

Other fasteners that you will find useful include lag screws, hanger bolts, carriage bolts, corrugated fasteners, and metal splines.

Screw washers are sometimes needed to provide added bearing surfaces and to avoid marring the wood. These washers come flat, countersunk, or flush. The choice will depend on the desired appearance and the type of material being fastened together.

OTHER METAL FASTENERS

New metal fasteners are constantly introduced into the market, so you need to consult your hardware supplier on regular basis to keep up with these new devices. An example of these new products is the fastener used for attaching items to hollow surfaces. These fasteners include gravity toggles, split-wing toggles, collapsible anchors, and hollow-door anchors. Each of these fasteners has a variety of applications and each is better suited than another for a specific use. You need to be familiar with them so you can make the proper choice when the job calls for their application.

METAL STUDS

Metal studs consist of metal channels with openings through which electrical and plumbing lines can be installed. These studs are fastened to base and ceiling channels with metal screws or clips. They come preformed and ready for installation, although they can be cut with a metal-cutting blade if necessary. Wallboard and other such wall materials are fastened to the studs with self-tapping drywall screws. Metal stud systems are usually designed for non-load-bearing walls and partitions.

These studs were originally designed for use in commercial and institutional construction, but now they are being used in residential settings. Their assembly continues to be simplified, so that they are quick to install and easy to remove when a temporary wall is needed.

8
Reading Blueprints

Carpenters must be able to read and understand architectural drawings and correctly interpret the information found in written specifications. The drawings and written specifications provide the plans for the design of the structure and for the required construction materials.

Copies made of original drawings are usually called blueprints. This name came from the old process of using chemically treated paper to create the plans. The lines were white on a blue background. In modern construction, the term "blueprint" still applies, although the drawings may have dark lines on white paper or some other combination. The use of computer-assisted drafting has brought a number of additions and modifications to traditional blueprints, but the process of reading and using them has not changed.

Building a structure is a very complicated operation that requires many different sets of information that can not be contained on one page. So a set of plans is created by making many different sheets of drawings and binding them together. These drawings help all those involved in the construction process to understand their responsibilities. A set of plans must be provided to the carpenter, building contractor, tradespersons, such as the electrician, plumber, and heating contractor, and the building owner. The plans are used by lumber dealers and suppliers to estimate the materials that will be needed.

A set of plans normally includes the following:

• a plot plan
• a foundation or basement plan
• floor plans
• elevation drawings (showing the front, rear, and sides of the building)
• drawings of the electrical, plumbing, heating, and air-conditioning layouts, commonly called mechanicals.

Buildings are reduced in scale when they are drawn so that they will fit onto the drafting sheet. Residential plan views are generally drawn to ¼-inch scale (¼″ = 1′). This means that for each ¼-inch on the plan, the building dimension will be 1 foot. The dimensions shown on the plan, however, are actual sizes.

When certain parts of the structure need to be shown in greater detail, they are drawn to a larger scale, such as 1″ = 1′. Other parts, such as framing, are often drawn to a smaller scale (⅛″ = 1′).

CHANGING PLANS

Minor changes in the plans occur throughout the construction process because of the desires of the owner or the identification of a better layout method. These changes often include changing the size and location of a window, or the design of a built-in cabinet or closet opening. You can generally handle these changes without causing any disruption in the construction process. You should include your changes with sketches and notations on the plans in order to eliminate any misunderstandings at a later date.

Major changes such as moving load-bearing walls or relocating stairs should be undertaken only after they are cleared with the architect, owner, and building contractor. This will avoid the possibility of a series of problems that could affect the soundness of the structure or installation of utilities.

DETAILS

Details are large-scale drawings that show features that do not appear (or appear in too small a scale) on the plans, elevations, and sections. Details do not have a cutting-plane indication, but are simply noted by code. The construction of doors, windows, and eaves is usually shown in detail drawings. Other details, which are customarily shown, are sills, girder and joist connections, and stairways.

ELEVATION DRAWINGS

Elevation drawings show the front, rear, or side view of a building or structure. Construction materials may be shown on the elevation

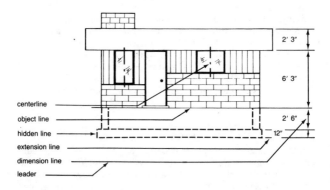

(Figure 8-1)   Alphabet of lines

drawing. The drawing may also show the ground level surrounding the structure, called the grade. When more than one view is shown on a drawing sheet, each view is given a title. If any view has a scale different from that shown in the title block, the scale is given beneath the title of that view.

The centerline symbol of alternate long and short dashes in an elevation drawing shows finished floor lines. Foundations below the grade line are shown by the hidden line symbol of short, evenly spaced dashes. (See Figure 8–1.) Elevation drawings also show the locations and types of doors and windows.

FOUNDATION AND FLOOR PLANS

Foundation plans are often combined with the basement plans to show where the footings are to be set and the layout of the basement floor.

Floor plans show the size and outline of the building and its rooms. They include the outside shape of the building; the arrangement, size, and shape of the rooms; the type of materials; and the length, thickness, and character of the building walls at a particular floor. A floor plan also includes the types and locations of utility

Symbol	Width	Height	Thickness	Material	Type	Screen	Quantity	Threshold	Finish	Manufacturer
Ⓐ	3' 0"	6' 8"	1-3/4"	Ash	Solid	Aluminum	1	Aluminum	Ext. Varnish	Ace Door Co.
Ⓑ	2' 8"	6' 8"	1-3/4"	Ash	Slab Core	Self-Storing Storm Door	1	Aluminum	Ext. Varnish	Ace Door Co.
Ⓒ	2' 6"	6' 8"	1-3/8"	Ash	Hollow Core	No	1	No	Int. Varnish	-
Ⓓ	2' 8"	6' 8"	1-3/8"	Ash	Hollow Core	No	4	No	Int. Varnish	-
Ⓔ	2' 4"	6' 8"	1-3/8"	Ash	Hollow Core	No	3	No	Int. Varnish	-

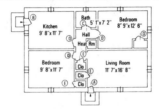

(Figure 8-2a)   Door schedule

installations, and the location of stairways, as well as the type, width, and location of the doors and windows. Door and window schedules may accompany the floor plans to provide additional details. (See Figure 8–2 a and b.)

PLOT PLANS

Plot plans (also called site plans) show all of the property lines, contours, and profiles; building lines, including the location of structures to be constructed as well as existing structures; approaches; finished grades; and existing and new utilities, such as sewer, water, and gas.

The site plan has a north-pointing arrow to indicate site north. Each facility has a number (or code letter) to identify it in the schedule of facilities. The contour lines show the elevation of the earth surfaces; all points on a contour have the same elevation. Distances are given between principal details and reference lines.

Symbol	Width	Height	Material	Type	Screen	Quantity	Remarks	Manufacturer	Catalog #
⚠1	5′ 0″	4′ 0″	Aluminum	Stationary	No	1		A&S Glass	548W
⚠2	4′ 0″	4′ 0″	-	Double Hung	Yes	7		A&S Glass	440W
⚠3	3′ 0″	3′ 0″	-	-	-	1	Frosted Glass	A&S Glass	330W

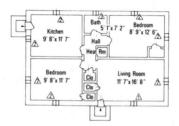

(Figure 8-2b)   Window schedule

All distances in a plan view give the horizontal measurement between two points and do not show terrain irregularities.

SECTION DRAWINGS

Section drawings show how a structure looks when cut vertically by a cutting plane. It is drawn to a large scale showing details of a particular construction feature that cannot be given in the general drawing. The section drawing provides information on height, materials, fastening and support systems, and concealed features. Parts of the structure likely to have a section drawing include walls, window and door frames, footings, and foundations. Wall section drawings, for example, show not only the construction of the wall but also how the structural members are joined together and how they tie together from the foundation bed to the roof.

SPECIFICATIONS

Specifications (or specs) are often given in written form and include the following information:

- Basic information for general requirements and conditions
- excavating and grading
- masonry and concrete work
- rough carpentry
- roofing
- finish carpentry
- insulation
- drywall
- tile work
- painting and finishing for interior and exterior
- caulking and sealing
- sheet metal work
- electrical work
- plumbing
- heating and air-conditioning
- landscaping

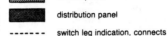 service panel

distribution panel

- - - - - - switch leg indication, connects outlets with control points

— ·· — ·· — low-voltage relay system wiring

(T) bell-ringing transformer

(J) junction box

bell

 GR duplex convenience outlet for grounding-type plugs

 WP weatherproof convenience outlet

 S combination switch and convenience outlet

(Figure 8-3a) Electrical symbols

Under each heading, a scope of work is given that includes kinds of materials to be used, methods of application, and level of quality that is to be delivered.

You should always check the specifications carefully. They are invaluable as you plan your work and material selection, and they help ensure that the building owner will be satisfied that a quality job has been completed.

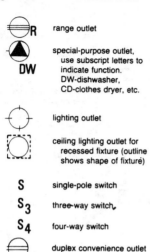

range outlet	
special-purpose outlet, use subscript letters to indicate function. DW-dishwasher, CD-clothes dryer, etc.	
lighting outlet	
ceiling lighting outlet for recessed fixture (outline shows shape of fixture)	
S	single-pole switch
S₃	three-way switch
S₄	four-way switch
	duplex convenience outlet

SYMBOLS

Architectural plans are drawn to scale, and thus there is not enough room on the plans to include drawings for all of the details, such as those for electrical and plumbing fixtures. So, architects use symbols and conventional representations to indicate types of materials and other items. (See Figure 8–3(a) and (b)). These simplify the illustration of assemblies and other elements of the structure.

(Figure 8-3a) Electrical symbols (cont'd)

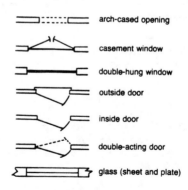

arch-cased opening

casement window

double-hung window

outside door

inside door

double-acting door

glass (sheet and plate)

(Figure 8-3b) Architectural symbols

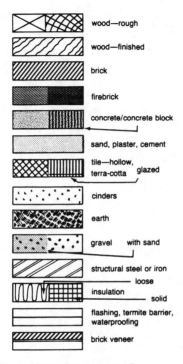

(Figure 8-3b)   Architectural symbols (cont'd)

PART FOUR

APPENDICES

Appendix 1
Job and
Workplace Safety

Safety on the job site is a combination of common sense, concern, and caution. Safety awareness must become a part of your daily work habits, and safety procedures and prescriptions must be strictly followed. Specific safety tips are included throughout the book and should be particularly noted in the chapters describing the use of hand tools and power tools. This appendix presents some general guidelines and practices that will contribute to a healthy work environment.

GENERAL GUIDELINES

ATTITUDE

Everybody has good and bad days on the job, but if you are working under a tight deadline or stress, the chances of an accident occurring are increased. Try to slow the process down to the point that you are not taking unnecessary risks or shortcuts.

BENDING AND LIFTING

Back injuries can be avoided by properly lifting or carrying heavy objects. When lifting a heavy object, stand close to it, bend your

291

knees, and grasp the object firmly. Then lift by straightening your legs and keeping your body as nearly vertical as possible. To lower the object, reverse the procedure. When carrying a heavy load, do not turn or twist your body but make adjustments in position by shifting your body. If the load is heavy or bulky, secure help from others—don't try to do everything yourself.

CLOTHING

Wear clothing that is appropriate for the work and weather. Overalls, coveralls, or pant legs should not be loose or floppy. All pants should be without cuffs, so that objects cannot fall into them and so that they cannot become entangled in a piece of equipment or material. Shirtsleeves should be buttoned or rolled up. Never wear loose or floppy clothing around moving machinery.

Shoes should be sturdy with thick soles. Lightweight sports-type shoes or leather-sole shoes should not be worn. Sports shoes do not give protection from falling objects, nor do they supply adequate support over a long workday; leather soles are very slick and do not provide proper traction on smooth wood surfaces such as roof sheathing. On heavy construction jobs, safety boots or shoes are required. They have steel reinforced toes that withstand crushing and prevent injured toes.

A hard hat is needed on the construction site to prevent head injury. The hard hat selected should meet standard specifications in order for it to provide maximum protection.

Appropriate gloves should be worn. Some gloves are designed to protect the hands from rough materials; others are designed to protect the hands from chemical reactions. Make sure the gloves provide the maximum protection for the job being undertaken.

EXCAVATIONS

When working below grade, such as in basement areas, make sure there is adequate shoring to keep the soil in place. Cave-ins can occur at anytime, but are especially apt to occur after a rain or when other excavation is being done and the heavy equipment causes vibration in the surrounding soil. When going into and out of excavations, use a ladder or steps so that you do not have to put pressure on the soil walls. Also, check whether there are any underground utilities in the excavation area so that you do not cut into them.

FALLING OBJECTS

Always be on the lookout for falling objects, and make sure that you do not contribute to the possibility of objects falling. Do not place tools on the edge of scaffolds, stepladders, window sills, or on any other surface where they might be knocked off. When moving long pieces of lumber, make sure you know where both ends are at all times. If you place lumber so that it leans against the side of the structure, make sure it is secure. Don't leave it there for any length of time.

FIRST AID

You should have a knowledge of first aid. You should understand approved procedures and be able to exercise good judgment in applying them. An accident victim may receive additional injury from unskilled treatment by an unqualified person. Make sure you know what you are doing. A Red Cross training program can provide you with the skill you will need to deal with an emergency. Always keep a first-aid kit on the job site so that any minor injuries can be cleaned, sterilized, and bandaged as soon as possible in order to reduce the chances of infection.

HOUSEKEEPING

Always keep the work site clean. Scraps in the work area can cause a fall, and result in injury. Building materials should be placed in neat piles with adequate aisles between the piles for easy access. Tools and equipment should be stored in panels or chests so that they will not interfere with the work area and will be protected when not in use.

Always keep the work surface clear of scraps and rubbish. When working on roof decks, make sure that any snow or ice is removed. Do not climb on smooth surfaces if they are wet. Always be aware of any openings, such as stairwells, in the work area.

SOURCES OF INFORMATION

Workers' compensation carriers and other insurance companies can provide information on safety standards. Trade associations can also be helpful. If you belong to one, find out what the local organization does to assist members. If you are not a member, find out whether these groups provide information to nonmembers (many do).

The National Safety Council (NSC) is a nonprofit organization that provides a broad range of information about workplace safety and health. Check to see whether a local chapter of the NSC exists in your area, and find out what materials are available that pertain to your situation. If no local chapter is nearby, write to the national office at:

> National Safety Council
> 425 North Michigan Avenue
> Chicago, IL 60611

LEGAL SAFETY STANDARDS

The Occupational Safety and Health Act of 1970 made safety and health on the job a matter of law for all businesses and their employees. The Occupational Safety and Health Administration (OSHA) functions to promote and implement safety and health standards, issue regulations, provide training programs for employers and employees, improve substandard working conditions, and assist in establishing plans and programs that will be in compliance with legal regulations.

In addition, the U.S. Secretary of Labor is empowered to establish standards necessary to provide for a safe and healthy workplace. Two sets of standards that you should be aware of are the General Industry Standards (29 CFR 1910) and Construction Standards (29 CFR 1926). It is important to obtain a current copy of those standards that apply to your situation.

Appendix 2
Codes and Regulations

Building codes and regulations are issued by various agencies to set minimum standards of safety in the design and construction of buildings. Building codes regulate the types of construction, function of the structure (called occupancy), quality of materials used in construction, the allowable loads and stresses for the structure, and types and uses of mechanical and electrical equipment. Codes and regulations impose other requirements related to buildings, with special emphasis on fire safety.

The codes issued by the different agencies are quite similar. However, there may be significant differences, and so the code adopted by a given municipality must be satisfied by buildings constructed within its jurisdiction.

MODEL BUILDING CODES

Model building codes are created and used on a national basis. They are most often used when municipalities have not generated their own. These model codes are based on generally accepted standards, which can be modified by the local municipalities to meet local needs.

The current trend in developing standards that are incorporated in model codes is to create what is called a performance code. This type of code requires a specified result. The advantage of this approach is that it permits the use of newer materials and methods of construction. The older approach was rigid in the specification of traditional materials.

Many states have established building codes that may control all building construction within the state; however, individual communities may adopt their own codes. An example is the State Building Construction Code adopted by the State of New York. As stated therein: "The municipalities of the state have the option to accept or not to accept the applicability" of this code.

Several regional agencies have formulated model building codes. These have been prepared to assist municipalities in preparing local codes. However, a municipality may adopt such a code, by reference, if authorized by the statutes of the governing state.

Check the building codes of the community in which you are building when you obtain the building permit. Also, if you have any questions, you should clear them through the city or county engineer's office and the building inspector that has been assigned to your project. These inspectors enforce the building codes by conducting on-site inspections throughout the construction process. They are generally individuals who have worked in the construction field and have a thorough understanding of the codes and their application.

After each inspection, the building inspectors fill out an inspection card that gives their approval for the work that has been completed to date. If they reject some work, they fill out the card when the problem has been corrected. This card should be kept with the building permit so that if any question is asked about work that has been inspected and approved, the card can be shown for verification. Work on the job should never go beyond the point indicated at the last inspection. You must pay close attention to the inspection card so that you do not enclose an area that has not been inspected. This is especially true when boxing in mechanical work (heating and plumbing) and electrical wiring. If these elements are enclosed before being inspected, the work will have to be torn out to the point that the inspector can clearly see the area that must be inspected.

BUILDING PERMITS

Building permits are required in almost every setting, and you should make sure the permit is in proper order before starting the job. Permits generally are secured by the owner or contractor by filing a formal application with the county or city clerk. The information required on the application includes a description of the job and two sets of plans. The plans submitted must include floor plans, specifications, site plan, and elevation drawings.

Generally, a filing fee and a plan review fee are required. In addition, a building permit fee is required, which is based on the cost of construction. When the building permit has been secured, it is posted on the construction site along with the inspection card. Some cities combine the permit and the inspection card on one document, and it is posted for easy observation.

WOOD CLASSIFICATION

Name of Wood	Classification	Hardness	Weight	Grain	Color	Workability
Pine, sugar white	coniferous	soft (s)	light (l)	closed (c)	near yellow	excellent
Pine, yellow	coniferous	medium (m)	medium (m)	closed (c)	yellow	poor (splits and has too much sap)
Poplar, yellow	deciduous	soft (s)	light (l)	closed (c)	yellow to white	good
Basswood	deciduous	soft (s)	light (l)	closed (c)	white	good
Fir	coniferous	medium (m)	light (l)	closed (c)	light brown	fair (splits)
Gum, red	deciduous	medium (m)	medium (m)	open (o)	reddish-brown	good
Cedar, red	coniferous	medium (m)	medium (m)	closed (c)	white, red	poor (splits and has knots)
Walnut	deciduous	hard (h)	medium (m)	open (o)	brown	hard
Mahogany	deciduous	hard (h)	medium (m)	open (o)	reddish-brown	good
Cherry	deciduous	hard (h)	medium (m)	closed (c)	light brown	hard
Maple	deciduous	hard (h)	heavy	closed (c)	light brown	hard

WOOD PANELING
Coverage Estimator

The following estimator provides factors for determining the exact amount of material needed for the five basic types of wood paneling. Multiply square footage to be covered by factor (length × width × factor).*

	Nominal Size	WIDTH Overall	WIDTH Face	AREA FACTOR
SHIPLAP	1 × 6	5½	5⅛	1.17
	1 × 8	7¼	6⅞	1.16
	1 × 10	9¼	8⅞	1.13
	1 × 12	11¼	10⅞	1.10
TONGUE AND GROOVE	1 × 4	3⅜	3⅛	1.28
	1 × 6	5⅜	5⅛	1.17
	1 × 8	7⅜	6⅞	1.16
	1 × 10	9⅜	8⅞	1.13
	1 × 12	11⅜	10⅞	1.10
S4S	1 × 4	3½	3½	1.14
	1 × 6	5½	5½	1.09
	1 × 8	7¼	7¼	1.10
	1 × 10	9¼	9¼	1.08
	1 × 12	11¼	11¼	1.07

	Nominal Size	WIDTH Overall	WIDTH Face	AREA FACTOR
PANELING PATTERNS	1 × 6	5⁷⁄₁₆	5⁵⁄₁₆	1.19
	1 × 8	7⅛	6¾	1.19
	1 × 10	9⅜	8¾	1.14
	1 × 12	11⅛	10¾	1.12
BEVEL SIDING	1 × 4	3½	3½	1.60
	1 × 6	5½	5½	1.33
	1 × 8	7¼	7¼	1.28
	1 × 10	9¼	9¼	1.21
	1 × 12	11¼	11¼	1.17

*Allowance for trim and waste should be added.

COURTESY OF WESTERN WOOD PRODUCTS ASSOCIATION

LUMBER GRADES
Grade-Use Guide for Appearance Grades of Plywood[1]

Use these symbols when you specify plywood	Description and Most Common Uses	Veneer Grade			Most Common Thickness (inch) (3)				
N-N, N-A, N-B INT-DFPA	Cabinet quality. One or both sides select all heartwood or all sapwood veneer. For natural finish furniture, cabinet doors, built-ins, etc. Special order items.	N	N,A, or B	C					¾
N-D-INT-DFPA	For natural finish paneling. Special order item.	N	D	¼					
A-A INT-DFPA	For interior applications where both sides will be on view. Built-ins, cabinets, furniture and partitions. Face is smooth and suitable for painting.	A	A	¼	⅜	½	⅝		¾
A-B INT-DFPA	For uses similar to interior A-A but where the appearance of one side is less important and two smooth solid surfaces are necessary.	A	B	¼	⅜	½	⅝		¾
A-D INT-DFPA	For interior uses where the appearance of only one side is important. Paneling, built-ins, shelving, partitions and flow racks.	A	D	¼	⅜	½	⅝		¾
B-B INT-DFPA	Interior utility panel used where two smooth sides are desired. Permits circular plugs. Paintable.	B	B	¼	⅜	½	⅝		¾

Interior

LUMBER GRADES—Continued

Grade-Use Guide for Appearance Grades of Plywood[1]

	Use these symbols when you specify plywood	Description and Most Common Uses	Veneer Grade			Most Common Thickness (inch) (3)					
Interior	B-D INT-DFPA	Interior utility panel for use where one smooth side is required. Good for backing, sides of built-ins, industry, shelving, slip sheets, separator boards and bins.	B	D	D	1/4		3/8	1/2	5/8	3/4
	DECORATIVE PANELS	Rough-sawn, brushed, grooved or striated faces. Good for paneling, interior accent walls, built-ins, counter facing, displays and exhibits.	C or btr.	D	D		5/16	3/8	1/2	5/8	
	PLYRON INT-DFPA	Hardboard face on both sides. For counter tops, shelving, cabinet doors, flooring. Hardboard faces may be tempered, untempered, smooth or screened.			C & D				1/2	5/8	3/4
Exterior	A-A EXT-DFPA (4)	Use in applications where the appearance of both sides is important. Fences, built-ins, signs, boats, cabinets, commercial refrigerators, shipping containers, tote boxes, tanks, and ducts.	A	A	C	1/4		3/8	1/2	5/8	3/4
	A-B EXT-DFPA (4)	For use similar to A-A EXT panels but where the appearance of one side is less important.	A	B	C	1/4		3/8	1/2	5/8	3/4

Exterior

Grade	Description	Face	Back	1/4 or 5/16	3/8	1/2	5/8	3/4
A-C EXT-DFPA (4)	Exterior use where the appearance of only one side is important. Sidings, soffits, fences, structural uses, boxcar and truck lining and farm buildings. Tanks, trays, commercial refrigerators.	A	C	1/4	3/8	1/2	5/8	3/4
B-B EXT-DFPA (4)	An outdoor utility panel with solid paintable faces.	B	B	1/4	3/8	1/2	5/8	3/4
B-C EXT-DFPA (4)	An outdoor utility panel for farm service and work buildings, boxcar and truck lining, containers, tanks, agricultural equipment.	B	C	1/4	3/8	1/2	5/8	3/4
HDO EXT-DFPA (4)	Exterior type High Density Overlay plywood with hard, semi-opaque resin-fiber overlay. Abrasion resistant. Painting not ordinarily required. For concrete forms, cabinets, counter tops, signs and tanks.	A or B	C plgd	5/16	3/8	1/2	5/8	3/4
MDO EXT-DFPA (4)	Exterior type Medium Density Overlay with smooth, opaque, resin-fiber overlay heat-fused to one or both panel faces. Ideal base for paint. Highly recommended for siding and other outdoor applications. Also good for built-ins, signs and displays.	B or C	C (5)	5/16	3/8	1/2	5/8	3/4
303 SIDING EXT-DFPA (7)	Grade designation covers proprietary plywood products for exterior siding, fencing, etc., with special surface treatment such as V-groove, channel groove, striated, brushed, rough-sawn.	(6)	C		3/8	1/2	5/8	

LUMBER GRADES—Continued
Grade-Use Guide for Appearance Grades of Plywood[1]

Use these symbols when you specify plywood	Description and Most Common Uses	Veneer Grade		Most Common Thickness (inch) [3]				
Exterior								
T 1-11 EXT-DFPA	Exterior type, sanded or unsanded, shiplapped edges with parallel grooves 1/4" deep, 3/8" wide. Grooves 2" or 4" o.c. Available in 8' and 10' lengths and MD Overlay. For siding and accent paneling.	C or btr.	C				5/8	
PLYRON EXT-DFPA	Exterior panel surfaced both sides with hardboard for use in exterior applications. Faces are tempered, smooth or screened.		C			1/2	5/8	3/4
MARINE EXT-DFPA	Exterior type plywood made only with Douglas fir or Western larch. Special solid jointed core construction. Subject to special limitations on core gaps and number of face repairs. Ideal for boat hulls. Also available with overlaid faces.	A or B	B	1/4	3/8	1/2	5/8	3/4

(1) SANDED BOTH SIDES EXCEPT WHERE DECORATIVE OR OTHER SURFACES SPECIFIED.

(2) AVAILABLE IN GROUP 1, 2, 3, 4, OR 5 UNLESS OTHERWISE NOTED.

(3) STANDARD 4×8 PANEL SIZES, OTHER SIZES AVAILABLE.

(4) ALSO AVAILABLE IN STRUCTURAL I (FACE, BACK AND INNER PLYS LIMITED TO GROUP 1 SPECIES).

(5) OR C-PLUGGED.

(6) C OR BETTER FOR 5 PLYS; C-PLUGGED OR BETTER FOR 3 PLY PANELS.

(7) STUD SPACING IS SHOWN ON GRADE STAMP.

COURTESY OF THE AMERICAN PLYWOOD ASSOCIATION.

ADHESIVE

• Best product for this use ○ Acceptable for this use **Adhesive** — **Properties**	Wood to wood or plywood	Wood veneering	Plastic laminates to wood	Wood boats, marine uses	Wood for outdoor use
Casein — Powered glue; mix with water; water resistant.	○	○	○		
Contact cement — Liquid, ready-to-use neoprene-based adhesive; no clamping needed; water resistant.		●	○		
Epoxy cement — Two-part resin and hardener; bonds almost anything.					
Household cement — Liquid, clear and fast-drying (5–15 minutes); vinyl-base; water resistant.					
Liquid hide glue — Liquid, ready-to-use brown animal glue; reliable; clamping time 40–50 minutes.	○	○	●		
Liquid solder — Liquid; requires no heat or mixing; water resistant.					
Plastic resin — Powdered, urea-based; mixes easily with water; good water resistance; requires 6 to 8 hours clamping.	○	○	○		○
Yellow glue — Liquid, ready-to-use aliphatic resin. Tacky, highest strength; good heat resistance; durable, tough bond.	●	○	○		
Resorcinol resin — Two-part powder and catalyst mix; fully waterproof; requires overnight clamping.	○	○	○	●	●
White glue — Liquid, ready-to-use polyvinyl acetate resin; sets fast (20–30 minutes clamping); dries clear.	○	○	○		

Metal to wood	China repair	Patch seal solder	Metal to metal	Paper to paper or cloth	Leather to leather or wood	Rubber to wood or metal	Cloth to cloth or wood	Plywood panels to frame or stud
			○		○	●	○	●
●	○	○	●					
	●			○			○	
				○	●		○	
		●						
				○	○		○	
				●	○		●	

SPAN TABLES
Floor Joist Span Tables (Feet and Inches)

Design Criteria:
Strength—10 lbs. per sq. ft. dead load plus 40 lbs. per sq. ft. live load.
Deflection—Limited to span in inches divided by 360 for live load only.

40# Live Load
10# Dead Load

Species or Group	Grade*	2 × 6			2 × 8			2 × 10			2 × 12		
		12" oc	16" oc	24" oc	12" oc	16" oc	24" oc	12" oc	16" oc	24" oc	12" oc	16" oc	24" oc
DOUGLAS FIR-LARCH	2	10-11	9-11	8-6	14-4	13-1	11-3	18-4	16-9	14-5	22-4	20-4	17-6
	3	9-3	8-0	6-6	12-2	10-7	8-8	15-7	13-6	11-0	18-11	16-5	13-5
DOUGLAS FIR SOUTH	2	10-0	9-1	7-11	13-2	12-0	10-6	16-9	15-3	13-4	20-5	18-7	16-3
	3	9-0	7-9	6-4	11-9	10-3	8-4	15-1	13-1	10-8	18-4	15-11	13-0
HEM-FIR	2	10-3	9-4	7-7	13-6	12-3	10-0	17-3	15-8	12-10	20-11	19-1	15-7
	3	8-3	7-2	5-10	10-10	9-5	7-8	13-10	12-0	9-10	16-10	14-7	11-11
MOUNTAIN HEMLOCK-HEM-FIR	2	9-5	8-7	7-6	12-5	11-4	9-11	15-11	14-6	12-8	19-4	17-7	15-4
	3	8-3	7-2	5-10	10-10	9-5	7-8	13-10	12-0	9-10	16-10	14-7	11-11
WESTERN HEMLOCK	2	10-3	9-4	7-11	13-6	12-3	10-6	17-3	15-8	13-4	20-11	19-1	16-3
	3	8-8	7-6	6-1	11-5	9-11	8-1	14-7	12-8	10-4	17-9	15-5	12-7
ENGELMANN SPRUCE-LODGEPOLE PINE (Engelmann Spruce-Alpine Fir)	2	9-5	8-7	6-11	12-5	11-2	9-1	15-11	14-3	11-7	19-4	17-3	14-2
	3	7-5	6-5	5-3	9-9	8-6	6-11	12-6	10-10	8-10	15-3	3-2	10-9
LODGEPOLE PINE	2	9-8	8-10	7-3	12-10	11-8	9-7	16-4	14-11	12-3	19-10	18-1	14-11
	3	7-10	6-10	5-7	10-5	9-1	7-5	13-4	11-7	9-5	16-3	14-1	11-6

SPAN TABLES

Floor Joist Span Tables (Feet and Inches)

Design Criteria:
Strength–10 lbs. per sq. ft. dead load plus 40 lbs. per sq. ft. live load.
Deflection–Limited to span in inches divided by 360 for live load only.

40# Live Load
10# Dead Load

Species or Group	Grade*	Span (feet and inches)											
		2 × 6			2 × 8			2 × 10			2 × 12		
		12" oc	16" oc	24" oc	12" oc	16" oc	24" oc	12" oc	16" oc	24" oc	12" oc	16" oc	24" oc
PONDEROSA PINE-LODGEPOLE PINE	2	9-5	8-7	7-0	12-5	11-4	9-3	15-11	14-5	11-9	19-4	17-7	14-4
	3	7-7	6-6	5-4	10-0	8-8	7-1	12-9	11-1	9-1	15-7	13-6	11-0
WESTERN CEDARS	2	9-2	8-4	7-3	12-0	11-0	9-7	15-4	14-0	11-6	18-9	17-0	14-11
	3	7-10	6-10	5-6	10-5	9-1	7-5	13-4	11-6	9-5	16-3	14-0	11-6
WHITE WOODS (Western Woods)	2	9-2	8-4	6-10	12-0	11-0	9-0	15-5	14-0	11-6	18-9	17-0	14-0
	3	7-5	6-5	5-3	10-5	8-6	6-11	12-6	10-10	8-10	15-3	13-2	10-9

*SPANS WERE COMPUTED FOR COMMONLY MARKETED GRADES. SPANS FOR OTHER GRADES CAN BE COMPUTED UTILIZING THE WWPA SPAN COMPUTER.
COURTESY OF WESTERN WOOD PRODUCTS ASSOCIATION

DRILL BITS

Drill and Auger Bit Sizes for Wood Screws

Screw size No.		1	2	3	4	5	6	7	8	9	10	12	14	16	18
Nominal screw Body diameter		.073	.086	.099	.112	.125	.138	.151	.164	.177	.190	.216	.242	.268	.294
Pilot hole	Drill size	$\frac{5}{64}$	$\frac{3}{32}$	$\frac{3}{32}$	$\frac{7}{64}$	$\frac{1}{8}$	$\frac{9}{64}$	$\frac{5}{32}$	$\frac{11}{64}$	$\frac{11}{64}$	$\frac{3}{16}$	$\frac{7}{32}$	$\frac{15}{64}$	$\frac{17}{64}$	$\frac{19}{64}$
	Bit size	—	—	—	—	—	—	—	—	—	—	—	4	5	5
Starter hole	Drill size	—	$\frac{1}{16}$	$\frac{1}{16}$	—	$\frac{5}{64}$	$\frac{5}{64}$	$\frac{3}{32}$	$\frac{7}{64}$	$\frac{1}{8}$	$\frac{1}{8}$	$\frac{9}{64}$	$\frac{5}{32}$	$\frac{3}{16}$	$\frac{13}{64}$
	Bit size	—	—	—	—	—	—	—	—	—	—	—	—	—	4

COURTESY OF U.S. GOVERNMENT PRINTING OFFICE

LUMBER DIMENSIONS

Standard Lumber Sizes/Nominal and Dressed, Based on WWPA Rules

Product	Description	Nominal Size		Dressed Dimensions		
				Thickness and Width		
		Thickness	Width	Surfaced Dry	Surfaced Unseasoned	Lengths
Dimension	S4S	2	2	1½	1 9/16	6′ and longer in multiples of 1′
		3	3	2½	2 9/16	
	Other surface combinations available.	4	4	3½	3 9/16	
			5	4½	4⅝	
			6	5½	5⅝	
			8	7¼	7½	
			10	9¼	9½	
			12	11¼	11½	
			Over 12	Off ¾	Off ½	
Scaffold Plank	S4S or Rough Full Sawn	1¼ and thicker	8 and wider	If dressed refer to "Dimension" sizes		6′ and longer in multiples of 1′
Timbers	S4S or Rough	5 and larger		½ Off Nominal (S4S)		6′ and longer in multiples of 1′
Decking	2″ (Single T&G)	2	5	1½	4	6′ and longer in multiples of 1′
			6		5	
			8		6¾	
			10		8¾	
			12		10¾	
	3″ and 4″ (Double T&G)	3	6	2½	5¼	
		4		3½		

		Nominal Size		Dressed Dimensions (Dry)		
		Thickness	Width	Thickness	Face Width	Lengths
Flooring	(D&M), (S2S & CM)	3/8 1/2 5/8 1 1 1/4 1 1/2	2 3 4 5 6	5/16 7/16 9/16 3/4 1 1 1/4	1 1/8 2 1/8 3 1/8 4 1/8 5 1/8	4' and longer in multiples of 1'
Ceiling and Partition	(S2S & CM)	3/8 1/2 5/8 3/4	3 4 5 6	5/16 7/16 9/16 11/16	2 1/8 3 1/8 4 1/8 5 1/8	4' and longer in multiples of 1'
Factory and Shop	S2S	1 (4/4) 1 1/4 (5/4) 1 1/2 (6/4) 1 3/4 (7/4) 2 (8/4) 2 1/2 (10/4) 3 (12/4) 4 (16/4)	5 and wider (except 4 and wider in 4/4 No 1 Shop and 4/4 No 2 Shop)	3/4 (4/4) 1 5/32 (5/4) 1 13/32 (6/4) 1 19/32 (7/4) 1 13/16 (8/4) 2 3/8 (10/4) 2 3/4 (12/4) 3 3/4 (16/4)	Usually sold random width	4' and longer in multiples of 1'

COURTESY OF WESTERN WOOD PRODUCTS ASSOCIATION

NAILS

Size, Type, and Use of Nails

Size	Lgth (in.)	Diam (in.)	Remarks	Where used
2d	1	.072	Small head	Finish work, shop work.
2d	1	.072	Large flathead	Small timber, wood shingles, lathes.
3d	1¼	.08	Small head	Finish work, shop work.
3d	1¼	.08	Large flathead	Small timber, wood shingles, lathes.
4d	1½	.098	Small head	Finish work, shop work.
4d	1½	.098	Large flathead	Small timber, lathes, shop work.
5d	1¾	.098	Small head	Finish work, shop work.
5d	1¾	.098	Large flathead	Small timber, lathes, shop work.
6d	2	.113	Small head	Finish work, casing, stops, etc., shop work.
6d	2	.113	Large flathead	Small timber, siding, sheathing, etc., shop work.
7d	2¼	.113	Small head	Casing, base, ceiling, stops, etc.
7d	2¼	.113	Large flathead	Sheathing, siding, subflooring, light framing.
8d	2½	.131	Small head	Casing, base, ceiling, wainscot, etc., shop work.
8d	2½	.131	Large flathead	Sheathing, siding, subflooring, light framing, shop work.
8d	1¼	.131	Extra-large flathead	Roll roofing, composition shingles.
9d	2¾	.131	Small head	Casing, base, ceiling, etc.
9d	2¾	.131	Large flathead	Sheathing, siding, subflooring, framing, shop work.
10d	3	.148	Small head	Casing, base, ceiling, etc., shop work.
10d	3	.148	Large flathead	Sheathing, siding, subflooring, framing, shop work.
12d	3¼	.148	Large flathead	Sheathing, subflooring, framing.
16d	3½	.162	Large flathead	Framing, bridges, etc.
20d	4	.192	Large flathead	Framing, bridges, etc.
30d	4½	.207	Large flathead	Heavy framing, bridges, etc.
40d	5	.225	Large flathead	Heavy framing, bridges, etc.
50d	5½	.244	Large flathead	Extra-heavy framing, bridges, etc.
60d	6	.262	Large flathead	Extra-heavy framing, bridges, etc.

¹ THIS CHART APPLIES TO WIRE NAILS, ALTHOUGH IT MAY BE USED TO DETERMINE THE LENGTH OF CUT NAILS.
COURTESY OF U.S. GOVERNMENT PRINTING OFFICE

SCREWS
Carriage Bolt Sizes

Lengths (inches)	Diameters (inches)			
	3/16, 1/4, 5/16, 3/8	7/16, 1/2	9/16, 5/8, 3/4	
3/4	x	—	—	—
1	x	x	—	—
1 1/4	x	x	x	—
1 1/2, 2, 2 1/2, etc., 9 1/2, 10 to 20.	x	x	x	x

COURTESY OF U.S. GOVERNMENT PRINTING OFFICE

Lag Screw Sizes

Lengths (inches)	Diameters (inches)			
	1/4	3/8, 7/16, 1/2	5/8, 3/4	7/8, 1
1	x	x	—	—
1 1/2	x	x	x	—
2, 2 1/2, 3, 3 1/2, etc., 7 1/2, 8 to 10.	x	x	x	x
11 to 12	—	x	x	x
13 to 16	—	—	x	x

COURTESY OF U.S. GOVERNMENT PRINTING OFFICE

ABRASIVES

Types of Abrasives and Uses

Abrasive	Source	Color	Use
Aluminum oxide	Manufactured	Gray-brown	Hand or machine sanding
Emery	Mined	Black	Polishing
Flint	Mined	White	Hand sanding
Garnet	Mined	Reddish-brown	For general-purpose hand sanding
Silicon carbide	Manufactured	Shiny black	Wet/dry wood and metal sanding
Tripoli	Mined	Red	Polishing
Whiting	Mined	Gray to white	Polishing

Grade Classification for Abrasive Paper

Flint	New Method Garnet, Silicon Carbide	Old Method
	16	4
	20	3½
Very Coarse	24	3
	30	2½
	36	2
	40	1½
Coarse	50	1
	60	1/2
Medium	80	1/0
	100	2/0
	120	3/0
	150	4/0
	180	5/0
Fine	220	6/0
	240	7/0
	280	8/0
	320	
Extra Fine	360	—
	400	10/0
Super Fine	500	—
	600	—

BOARD FEET
Rapid Calculation of Board Measure

Width	Thickness	Board feet
3"	1" or less	1/4 of the length
4"	1" or less	1/3 of the length
6"	1" or less	1/2 of the length
9"	1" or less	3/4 of the length
12"	1" or less	Same as the length
15"	1" or less	1¼ of the length

Screw Sizes and Dimensions

Length (in.)	0	1	2	3	4	5	6	7	8	9	10	11	12	13	14	15	16	17	18	20	22	24	26	28	30
¼		x	x	x																					
3/8	x	x	x	x	x																				
½		x	x	x	x	x	x																		
5/8			x	x	x	x	x	x																	
¾			x	x	x	x	x	x	x	x															
7/8				x	x	x	x	x	x	x															
1					x	x	x	x	x	x	x	x													
1¼						x	x	x	x	x	x	x	x	x											
1½							x	x	x	x	x	x	x	x	x										
1¾								x	x	x	x	x	x	x	x	x	x								
2									x	x	x	x	x	x	x	x	x	x	x						
2¼										x	x	x	x	x	x	x	x	x	x						
2½											x	x	x	x	x	x	x	x	x						
2¾												x	x	x	x	x	x	x	x						
3													x	x	x	x	x	x	x	x	x	x			
3½															x	x	x	x	x	x	x	x	x		
4																	x	x	x	x	x	x	x	x	
4½																			x	x	x	x	x	x	x
5																				x	x	x	x	x	x
6																						x	x	x	x

Gauge and diameter

	0	1	2	3	4	5	6	7	8	9	10	11
Steel wire gauge	17	15	14	13	12	11	10	9	8	7	6	5
Diameter (inches)	.054	.072	.080	.091	.105	.120	.135	.148	.162	.177	.192	.207

	12	13	14	15	16	17	18	20	22	24	26	28
Steel wire gauge	4½	4	3	2½	2	1	½	0	00	00½	000	0000
Diameter (inches)	.216	.225	.243	.253	.262	.283	—	.306	.331	—	.362	.393

COURTESY OF U.S. GOVERNMENT PRINTING OFFICE

COMPARISON OF COMMON WOODWORKING SAWS

Saw	Use	Style of Teeth	Points per Inch	Length	Sawing Angle
Crosscut	Cuts across the grain	point	8–10	20–26 inches	45°
Rip	Cuts with the grain	edge	6–8	20–26 inches	90•
Coping	Cuts curves	point	16–20	5-inch blade	N/A
Back	Cuts straight lines	point	14–18	8–14 inches	N/A
Miter	Cuts miter joints and ends	point	14–18	20–14 inches	N/A
Compass or Keyhole	Cuts large-radius curves through interior of workpiece	point	10	12–14 inches	90°

LUMBER DIMENSIONS
Nominal Sizes and Standard Sizes of Lumber

Nominal size (in.)	American standard (in.)
1 × 3	3/4 × 2½
1 × 4	3/4 × 3½
1 × 6	3/4 × 5½
1 × 8	3/4 × 7¼
1 × 10	3/4 × 9¼
1 × 12	3/4 × 11¼
2 × 4	1½ × 3½
2 × 6	1½ × 5½
2 × 8	1½ × 7¼
2 × 10	1½ × 9¼
2 × 12	1½ × 11¼
3 × 8	2½ × 7¼
3 × 10	2½ × 9¼
3 × 12	2½ × 11¼
4 × 12	3½ × 11¼
4 × 16	3½ × 15¼
6 × 12	5½ × 11½
6 × 16	5½ × 15½
6 × 18	5½ × 17½
8 × 16	7½ × 15½
8 × 20	7½ × 19½
8 × 24	7½ × 23½

BOARD FEET
Actual Length

Nominal size (in.)	Actual length in feet								
	8	10	12	14	16	18	20	22	24
1 × 2		1⅔	2	2⅓	2⅔	3	3½	3⅔	4
1 × 3		2½	3	3½	4	4½	5	5½	6
1 × 4	2¾	3⅓	4	4⅔	5⅓	6	6⅔	7⅓	8
1 × 5		4⅙	5	5⅚	6⅔	7½	8⅓	9⅙	10
1 × 6	4	5	6	7	8	9	10	11	12
1 × 7		5⅝	7	8⅙	9⅓	10½	11⅔	12⅚	14
1 × 8	5⅓	6⅔	8	9⅓	10⅔	12	13⅓	14⅔	16
1 × 10	6⅔	8⅓	10	11⅔	13⅓	15	16⅔	18⅓	20
1 × 12	8	10	12	14	16	18	20	22	24
1¼ × 4		4⅙	5	5⅚	6⅔	7½	8⅓	9⅙	10
1¼ × 6		6¼	7½	8¾	10	11¼	12½	13¾	15
1¼ × 8		8⅓	10	11⅔	13⅓	15	16⅔	18⅓	20
1¼ × 10		10 5/12	12½	14 7/12	16⅔	18¾	20⅚	22 11/12	25
1¼ × 12		12½	15	17½	20	22½	25	27½	30
1½ × 4	4	5	6	7	8	9	10	11	12
1½ × 6	6	7½	9	10½	12	13½	15	16½	18
1½ × 8	8	10	12	14	16	18	20	22	24
1½ × 10	10	12½	15	17½	20	22½	25	27½	30
1½ × 12	12	15	18	21	24	27	30	33	36
2 × 4	5⅓	6⅔	8	9⅓	10⅓	12	13⅓	14⅔	16
2 × 6	8	10	12	14	16	18	20	22	24
2 × 8	10⅔	13⅓	16	18⅔	21⅓	24	26⅔	29⅓	32
2 × 10	13⅓	16⅔	20	23⅓	26⅔	30	33⅓	36⅔	40
2 × 12	16	20	24	28	32	36	40	44	48
3 × 6	12	15	18	21	24	27	30	33	36
3 × 8	16	20	24	28	32	36	40	44	48
3 × 10	20	25	30	35	40	45	50	55	60
3 × 12	24	30	36	42	48	54	60	66	72
4 × 4	10⅔	13⅓	16	18⅔	21⅓	24	26⅔	29⅓	32
4 × 6	16	20	24	28	32	36	40	44	48
4 × 8	21⅓	26⅔	32	37⅓	42⅔	48	53⅓	58⅔	64
4 × 10	26⅔	33⅓	40	46⅔	53⅓	60	66⅔	78⅓	80
4 × 12	32	40	48	56	64	72	80	88	96

COURTESY OF U.S. GOVERNMENT PRINTING OFFICE

SPAN TABLES
Ceiling Joist Span Tables (Feet and Inches)

10# Live Load (No Storage)

5# Dead Load

Design Criteria:
Strength—5 lbs. per sq. ft. dead load plus 10 lbs. per sq. ft. live load. No storage above.
Deflection—Limited to span in inches divided by 240 for live load only.

Species or Group	Grade*	2 × 4 16" oc	2 × 4 24" oc	Grade*	2 × 6 16" oc	2 × 6 24" oc	2 × 8 16" oc	2 × 8 24" oc
DOUGLAS FIR-LARCH	STD	8-3	6-9	2	18-1	15-7	23-10	20-7
				3	14-8	11-11	19-4	15-9
DOUGLAS FIR SOUTH	STD	8-1	6-8	2	16-6	14-5	21-9	19-0
				3	14-2	11-7	18-9	15-3
HEM-FIR	STD	7-6	6-1	2	16-11	13-11	22-4	18-4
				3	13-1	10-8	17-2	14-0
MOUNTAIN HEMLOCK-HEM-FIR	STD	7-8	6-2	2	15-7	13-8	20-7	18-0
				3	13-1	10-8	17-2	14-0
WESTERN HEMLOCK	STD	7-10	6-4	2	16-11	14-6	22-4	19-1
				3	13-9	11-3	18-1	14-10
ENGELMANN SPRUCE-LODGEPOLE PINE (Engelmann Spruce-Alpine Fir)	STD	6-9	5-6	2	15-6	12-8	20-7	16-8
				3	11-9	9-7	15-6	12-8

LODGEPOLE PINE	STD	7-1	5-10	2	16-1	13-3	21-2	17-6
				3	12-7	10-7	16-7	13-6
PONDEROSA PINE-LODGEPOLE PINE	STD	6-9	5-6	2	15-7	12-10	20-7	16-10
				3	12-0	9-10	15-10	12-11
WESTERN CEDARS	STD	7-1	5-10	2	15-1	13-2	20-0	17-6
				3	12-7	10-3	16-7	13-7
WHITE WOODS (Western Woods)	STD	6-7	5-4	2	15-1	12-6	20-2	16-5
				3	11-9	9-8	15-6	12-8

*SPANS WERE COMPUTED FOR COMMONLY MARKETED GRADES. SPANS FOR OTHER GRADES CAN BE COMPUTED UTILIZING THE WWPA SPAN COMPUTER.
COURTESY OF WESTERN WOOD PRODUCTS ASSOCIATION

THERMAL CONDUCTIVITY

The relatively low thermal conductivity, or "k," of western softwoods provides a significant amount of insulation. "k" is the amount of heat (BTUs) transferred in one hour through one square foot of material one inch thick with a difference in temperature of one degree Farenheit.

The thermal conductivity of wood incrases with moisture content and density. The "k" values for the western woods are shown in the table below.

Species	"k"	R/In.
Douglas fir—larch	1.06	.94
Douglas fir South	.99	1.01
Hem-fir	.92	1.08
Mountain hemlock	.98	1.02
Alpine fir	.75	1.33
Englemann spruce	.80	1.25
Lodgepole Pine	.92	1.08
Poderosa pine	.89	1.12
Sugar pine	.89	1.12
Idaho white pine	.84	1.19
Western cedar	.75	1.33
Western hemlock	.99	1.01

COURTESY OF WESTERN WOOD PRODUCTS ASSOCIATION

Glossary

Abrasive Paper paper or cloth covered on one side with a grinding material and used for smoothing and polishing. Materials used for this purpose include crushed flint, garnet, emery, and corundum.

Acoustical Materials tile, plaster, and other materials that absorb sound waves. Generally applied to interior wall surfaces to reduce reverberation or reflection of sound.

Adhesive a substance capable of holding materials together by surface attachment. A general term that includes cements, mucilage, paste, mastic, and glue.

Aggregate granulated particles of different substances collected into a compound or conglomerate mass. Used in making concrete.

Air Conditioning control of the temperature, the humidity, and the movement and purity of air in a building.

Air-Dried wood seasoned by exposure to the atmosphere without artificial heat.

Alteration any change in the facilities, structural parts, or mechanical equipment of a building that does not increase the cubic content.

Anchor an iron of special form to fasten together timber or masonry.

Anchor Bolt a bolt that fastens columns, girders, or other members to concrete or masonry.

Annual Rings rings or layers of wood that represent one growth period of a tree. In cross section, the rings indicate the age of the tree.

Asphalt Shingles a type of composition shingle made of felt saturated with asphalt or tar pitch and surfaced with mineral granules.

Backfill the replacement of earth around foundations after excavation.

Backing the bevel on the top edge of a hip rafter that allows the roofing board to fit the top of the rafter without leaving a triangular space between it and the lower side of the roof covering.

Balloon Frame the lightest and most economical form of construction, in which the studding and corner posts are set up in continuous length from first-floor line or sill to the roof plate.

Baluster a small pillar or column used to support a rail.

Balustrade a series of balusters connected by a rail, generally used for porches, balconies, and the like.

Band a low flat molding.

Base the bottom of a column; the finish of a room at the junction of the walls and floor.

Batten (Cleat) a narrow strip of board used to fasten several pieces together.

Batter Board a temporary framework used to assist in locating the corners when laying a foundation.

Bay Window a rectangular, curved, or polygonal window or group of windows usually supported on a foundation extending beyond the main wall of a building.

Beam an inclusive term for horizontal members, including joists, girders, rafters, and purlins.

Bedding a filling of mortar, putty, or other substance used to secure a firm bearing.

Bevel to cut to an angle other than a right angle, such as at the edge of a board or a door.

Bevel Board a board used in the framing of a roof or stairway to lay out bevels.

Bid an offer to supply, at a specified price, materials, supplies, and equipment, or the entire structure or section of a structure.

Bird's Mouth a notch cut on the underside of a rafter to fit it to the top plate. A notch is not full if rafter butts flush with top plate rather than overhangs it. Also called a seat cut.

Blemish any defect, scar, or mark that tends to detract from the appearance of wood.

Board lumber less than 2 inches thick.

Board Foot the equivalent of a 1-inch thick board that has a total surface area of 1 square foot.

Bridging pieces fitted between floor joists to distribute the floor load.

Butt Joint squared ends or edges adjoining each other.

Camber a slight arch in a beam or other horizontal member that prevents it from bending into a downward or concave shape due to its weight or load.

Cantilever structure or part thereof extending beyond its support.

Casement a window in which the sash opens on hinges.

Casing the trim around a door or window opening, either outside or inside, or the finished lumber placed around a post or beam, etc.

Caulk to seal and waterproof cracks and joints, especially around windows and exterior door frames.

Chair Rail an interior molding applied along the wall of a room to prevent chairs from marring the wall.

Chamfer corner of a board beveled usually at a 45 degree angle. Two boards butt-jointed and with chamfered edges form a V-joint.

Checks splits or cracks in a board caused by seasoning.

Chord the principal member of a truss on either the top or bottom.

Clamp a mechanical device used to hold two or more pieces together.

Column a square, rectangular, or cylindrical support for roofs, ceilings, etc., composed of a base, shaft, and capital.

Concrete an artificial building material made by mixing cement and sand with gravel, stones, or other aggregate, and sufficient water to cause the cement to set and bind the entire mass.

Contact Cement neoprene rubber-based adhesive that bonds instantly upon contact of the parts being joined.

Convenience Outlet electrical outlet into which portable equipment, such as mixers, may be plugged.

Cornice the molded projection that finishes the top of the wall of a building.

Counterflashing strips of metal used to prevent water from entering the top edge of the vertical side of roof flashing; the metal strips allow for expansion and contraction without danger of breaking the flashing.

Cove Molding molding with a concave profile used primarily where two members meet at a right angle.

Cripple Jack a rafter that intersects neither the plate nor the ridge and is terminated at each end by hip and valley rafters.

Cripple Stud a stud used above or below a wall opening. Extends from the header above the opening to the top plate or from the rough sill to the sole plate.

Crown the top or crested part of lumber; the point of greatest strength, where fibers are less likely to give way.

Cupola a small structure installed on a roof. It provides ventilation for the attic.

Dado in woodworking, a groove cut across the grain of wood.

Dead Load the weight of the unmovable parts of a structure, such as the weight of the walls and roof on the floor of the structure.

Deadening construction intended to prevent the passage of sound.

Deck the flat portion of a roof or floor upon which is put some type of covering.

Diagonal inclined member of a truss or bracing system used for stiffening and wind bracing.

Dovetail Joint joint made by cutting pins in an edge in the shape of dovetails that fit between the dovetails of another edge.

Dimension Lumber wood that has been cut to uniformly accepted standards for the type of lumber being produced. It is commonly cut 2 inches thick and from 4 to 12 inches wide.

Drip the projection of a window sill to allow water to drain clear of the side of the house below it.

Drip Cap a molding placed over an exterior window or door to cause the water to run off, or drip, outside the frame of the structure.

Drywall paper coated gypsum sheets used to cover interior walls.

Dutch Door a door divided horizontally. The bottom can be closed and secured while the top is left open.

Eave the part of the roof that projects over the side wall.

Edge the part of a board that is measured to determine thickness.

Elevation a drawing showing the external upright parts of a building or structures within a building (e.g., kitchen cabinets or fireplace).

Enamel a type of paint that dries to a hard glossy, semiglossy, or flat finish.

Fascia the vertical member of a cornice or other finish; generally the board of the cornice to which the gutter is fastened.

Filler a piece used to fill the space between two surfaces.

Flashing the material and process used for making watertight the roof intersections and other exposed places on the outside of a structure.

Float a tool, usually made of wood, that is used to level, but not smooth, concrete.

Floor Plan drawing of the layout of the various components of the floor structure.

Flush arranged edge to edge so that adjacent surfaces are even or on the same plane.

Footing a base that is wider than the wall and upon which the weight of a structure is distributed.

Footing Form a wooden or steel structure placed around the footing area that will hold the concrete to the desired shape and size until cured.

Foundation the part of a building or wall that supports the superstructure.

Frame the surrounding or enclosing woodwork of windows, doors, etc., and the skeleton of a building.

Framing the unfinished structure of a building, including interior and exterior walls, floors, roof, and ceilings.

Furring narrow strips of board nailed or glued on the walls and ceilings to form a straight surface upon which tiles or other finish materials are laid.

Gable the vertical triangular end of a building from the eaves to the apex of the roof.

Gauge a tool used by carpenters to strike a line parallel to the edge of a board.

Gambrel a symmetrical roof with two different pitches or slopes on each side.

Girder a wooden or metal post used to support wall beams or joists.

Girt the horizontal member of the wall of a full or combination frame house that supports the floor joists or is flush with the top of the joists.

Grade the horizontal ground level of a building.

Groove a hollow channel into which a piece fits or slides. Two special types of grooves are the dado (a rectangular groove cut across the full width of a piece) and the housing (a groove cut at any angle with the grain and part-way across a piece).

Hanger a vertical-tension member that supports a load.

Header a short joist into which the common joists are framed, around or above an opening.

Headroom the clear space between the floor line and ceiling, as in a stairway.

Heel of a Rafter the end or foot that rests on the wall plate.

Hip Roof a roof that slopes up toward the center from all sides, requiring a hip rafter at each corner.

Insulation a material used in walls, floors, and ceilings for the purpose of reducing transmission of hot or cold air.

Interior the inside of a structure, generally associated with the finishing process of the structure.

Jack Rafter a short rafter framing between the wall plate; a hip rafter.

Jamb the side piece or post of an opening; sometimes applied to the door frame.

Joint the line at which materials meet.

Joists structural members that are of 2-inch or thicker dimension that support the floors or ceilings of a structure.

Kerf the cut made by a saw.

Knee Brace a corner brace fastened at an angle from a wall stud to a rafter; used to strengthen a wooden or steel frame to prevent angular movement.

Lap a method of joining two pieces by placing one piece over part of the other, as in a lap joint.

Lattice crossed wood, iron, plates, or bars.

Ledger Board the support for the second-floor joists of a balloon-frame house or the like.

Level a term describing the position of a line or plane when parallel to the surface of still water; an instrument or tool used to test horizontal and vertical surfaces.

Lintel a horizontal structural member spanning an opening and supporting a wall load.

Live Load the movable or temporary weight applied to a structure, including furniture, people, snow, and wind.

Lookout the end of a rafter or the construction that projects beyond the sides of a house to support the eaves.

Louver a finned or slatted device installed, usually in the peaks of gables and the tops of towers, to exclude rain and snow while allowing airflow.

Lumber the sawed parts of a log such as boards, planks, scantling, and timber.

Mansard Roof a roof with two slopes on all four sides; the lower slope is steep, while the upper is almost flat.

Member a single piece of a structure which is complete in itself.

Miter the joint formed when two abutting pieces meet at an angle.

Mortise a hole that receives a tenon or projection or which has been cut into or through a piece by a chisel; generally of rectangular shape.

Mullion the construction that divides a window opening to accommodate two or more windows.

Mutin the vertical member between two panels of the same piece of panel work; the vertical sash bars separating the different panels of glass.

Newel the principal post at the foot of a staircase; the central support of a winding flight of stairs.

Nosing the part of a stair tread that projects over the riser; any similar projection; the rounded edge of a board.

On Center (O.C.) measurement from the center of one building member to another. Commonly used in laying studs and joists.

Oriented Strand Board (OSB) building panels made of wood strands arranged in layers at right angles to each other and bonded together under pressure with glue.

Partition a permanent interior wall that divides a building into rooms.

Piers supports set independently of the main foundation.

Pitch inclination or slope, as for roofs or stairs; equivalent to the rise divided by the run.

Pitch Board a board sawed to the exact shape formed by the stair tread, riser, and slope of a stairway and used to lay out the carriage and stringers.

Plan a horizontal geometrical section drawing of a building showing the wall, doors, windows, stairs, chimneys, columns, etc.

Plank a wide piece of sawed timber, usually 1½ to 4½ inches thick and 6 inches or more wide.

Plate the top or bottom horizontal piece of a structural frame or roof (as in top plate, sill plate, sole plate).

Plate Cut the cut in a rafter that rests upon the plate; sometimes called the bird's mouth.

Plow to cut a groove running in the same direction as the grain of the wood, or to cut a dado.

Plumb Cut a cut made in the vertical plane; the vertical cut at the top end of a rafter.

Ply a layer or thickness of wood (as in plywood) or building or roofing paper (as in two-ply, three-ply, etc.).

Porch an ornamental entrance way.

Post a timber set on end to support a wall, girder, or other structural member.

Pounds Per Square Foot (psf) generally used in calculating weight loads on floor joists.

Pounds Per Square Inch (psi) generally used in measuring air or water pressure.

Purlin a timber supporting several rafters at one or more points or supporting the roof sheathing directly.

Rabbet in woodworking, a groove cut in the surface or along the edge of a board so as to receive another board. A rabbet has only two surfaces: a side and a bottom.

Rafters the beams that slope the ridge of a roof to the eaves and make up the main body of the roof's framework. Common rafters are the beams that run square with the plate and extend to the ridge; cripple rafters are the beams that cut between the valley and hip rafters; hip rafters are the beams that extend from the outside angle of the plate toward the apex of the roof; jack rafters are the beams that square with the plate and intersect the hip rafters; valley rafters are the beams that extend from an inside angle of the plate toward the ridge or centerline of the house.

Rails the horizontal members of a balustrade or panel work.

Rake the trim of a building extending in an oblique line.

Return the continuation of a molding or finish of any kind in a different direction.

Ridge the top edge or corner formed by the intersection of two roof surfaces.

Rise the vertical distance through which anything rises, as the rise of a roof or stair.

Riser the vertical board placed between two treads of a flight of stairs.

Roofing the material put on a roof to protect it from wind and water.

Rubble roughly broken quarry stone or broken-up concrete.

Run the length of the horizontal projection of a piece, such as a rafter, when in position.

Saddle Board the finish of the ridge of a pitch-roof house. Sometimes called a comb board.

Sash the framework that holds the glass in a window.

Sawing (Plain) the method of sawing lumber regardless of the grain. The log is simply squared and sawed to the desired thickness.

Scab a short piece of lumber used to splice or to prevent movement of two other pieces.

Scaffold or Staging a temporary structure or platform enabling work above ground level.

Scale a measurement used in a drawing that is proportionate to the structure or part of the structure reflected in the drawing, such as ½ inch in the drawing equals 1 foot in the actual structure.

Scantling lumber with a cross section ranging from 2 × 4 to 4 × 4 inches.

Scarf a joint placed between two pieces of wood which allows them to be spliced lengthwise.

Scotia a hollow molding used as a part of a cornice and often placed under the nosing of a stair tread.

Scribing the marking of a piece of wood to provide for the fitting of one of its surfaces to the irregular surface of another.

Seat Cut or Plate Cut the cut at the bottom end of a rafter to allow it to fit upon the plate. Also called a bird's mouth.

Seat of a Rafter the horizontal cut on the bottom end of a rafter that rests on the top of the plate.

Section a drawing of a sliced view of an object showing the style, arrangement, and proportions of its various parts.

Shakes imperfections in lumber caused during the growth of the timber by high winds or other unfavorable conditions. Also, split wood shingles.

Sheathing wall or roofing boards or sheets applied to the joists or rafters and upon which are laid roofing or siding materials.

Siding the outside finish material put over sheathing.

Sills the horizontal members of a house that either rest upon the masonry foundations or, in the absence of such, form the foundations. Also the bottom members of window or door frames.

Sizing the process of working material to the desired size.

Sleeper a timber laid on the ground to support a floor joist, or a furring strip laid on a concrete slab to provide a nailing surface for flooring material.

Span the distance between the bearings of an arch or post.

Specifications the written or printed directions regarding the details of a building or other construction.

Splice the joining of two similar members in a straight line.

Square a tool used by carpenters to obtain accuracy in laying out perpendicular lines.

Square a term applied to surface area coverage equivalent to 100 square feet, e.g., one square of shingles will cover 100 square feet of roof.

Stringer a long horizontal piece of lumber in which notches have been cut to support the treads for a staircase.

Stucco a fine, decorative plaster used to cover interior walls; a coarser plaster used for rough, outside wall coverings.

Stud an upright piece of lumber used in the framework of a building.

Studding the framework of a partition or the wall of a house; usually made up of 2 × 4s.

Subfloor a wooden floor that is laid over the floor joists and upon which the finished floor is laid.

Threshold the beveled piece over which the door swings.

Tie Beam a beam so situated as to tie the principal rafters of a roof together and prevent them from thrusting the plate out of line.

Timber lumber with a cross section of more than 4 x 6 inches, such as that used for posts, sills, and girders.

Tin Shingle a small piece of tin used in flashing and repairing a shingle roof.

Top Plate a piece of lumber that supports the ends of rafters.

Tread the horizontal part of a stair.

Trim outside or interior finished woodwork; the finish around openings.

Trimmer the beam or floor joist into which a header is framed.

Trimming putting the inside and outside finish and hardware on a building.

Truss a structural framework of triangular units used to support loads over a long span.

Valley the internal angle formed by the two slopes of a roof.

Verge Boards boards that serve as the finish for the eaves on the gable end of a building.

Vestibule the entrance to a house; usually enclosed.

Wainscoting matching boards or panel work that covers the lower portion of a wall.

Wash the slant upon a sill, capping, etc., to allow the water to run off easily.

Wind a term used to describe the surface of a board when twisted.

INDEX